Graphical Models and Causal Discovery with Python

Joe Suzuki

Graphical Models and Causal Discovery with Python

100 Exercises for Building Logic

Joe Suzuki
Graduate School of Engineering Sciences
Osaka University
Toyonaka, Osaka, Japan

ISBN 978-981-95-5307-5 ISBN 978-981-95-5308-2 (eBook)
https://doi.org/10.1007/978-981-95-5308-2

This Springer imprint is published by the registered company Springer Nature Singapore Pte Ltd.
The registered company address is: 152 Beach Road, #21-01/04 Gateway East, Singapore 189721, Singapore

Preface

Graphical models provide a highly attractive framework that captures complex relationships among random variables both visually and mathematically. By fusing probability theory with graph theory, methods such as Bayesian networks and Markov networks have been systematized to model real-world structures. With these models, we can go beyond merely grasping correlations and move toward *causal discovery*, which reveals causal structures among variables.

Causal discovery, the topic of this book, consists of two key steps: not only estimating the pattern of dependencies (the "skeleton") among variables, but also placing arrows to determine the *causal ordering* of variables. This problem had been difficult to treat within traditional statistical frameworks alone, but recent advances—kernel methods, the use of non-Gaussianity, information criteria, and Bayesian marginal likelihood—have enabled more flexible and theoretically coherent approaches.

One of the charms of graphical models is that they provide an intuitive sense of "structure" that cannot be obtained from formulas or code alone. You can draw them as figures, explain them in words, and manipulate them by hand to understand them—this visual and operational understanding brings the abstract world of probability and statistical modeling much closer. Standing at the intersection of statistics and machine learning, and bridging theory and practice, this field will undoubtedly play an increasingly important role.

My encounter with graphical models dates back to the early 1990s, before the term "machine learning" had become as common as it is today. While an assistant at Waseda University, I happened upon the proceedings of the UAI (Uncertainty in Artificial Intelligence) conference at an English-language technical bookstore along Meiji-dori near the School of Science and Engineering. I was immediately drawn into the refined discussions on how to learn the structure of Bayesian networks from data.

Around the same time, I was introduced to Jorma Rissanen's MDL (Minimum Description Length) principle and strongly resonated with the idea of "discovering the simple structure behind data." I promptly applied MDL to the problem of learning graph structures from samples and presented the results at UAI in 1993.

I have been told that this was among the earliest presentations by a Japanese researcher to introduce structural learning of graphical models to the international community.

Since then, I have been captivated by the potential of graphical models and have continued my research. Behind diagrams that may look deceptively simple at first glance lie a large number of mathematical challenges. Facing each of them and carefully untangling them has been the starting point of my career as a researcher and remains at the core of my work today.

This book tackles two major tasks in causal discovery: exploring dependency structures and exploring variable orders.

Causal discovery = exploring dependencies + exploring variable order

Given p variables and n samples, neither step alone suffices to draw the correct causal structure. If you simply feed your data into a tool, it will readily return something that looks like dependencies and an ordering. However, when the results differ from your expectations or run counter to intuition, unless you can explain *why* the algorithm behaved that way, you are left with mere outcomes. You will struggle to write a convincing paper or to persuade your supervisor or client.

This book focuses specifically on causal discovery within graphical models, incorporating modern techniques such as kernel methods and Bayesian approaches. For LiNGAM—the directed approach to exploring variable order devised by Professor Shohei Shimizu (Osaka University)—I have aimed to describe its essence clearly.

Furthermore, I have made extensive use of exercises from a graduate course taught in 2019 and materials from more than ten iterations of an online small-group seminar. As with my previous "100 Problems in the Mathematics of Machine Learning" series, the goal is not merely to memorize procedures but to think through "why it works," and to deliver a book that genuinely inspires.

The structure of this book is designed to deepen understanding step by step while balancing theory and implementation, intuition and rigor. The diagram below visually summarizes the relationships and roles of each chapter.

This is not a comprehensive text on all of graphical models; rather, it is a specialized book that delves deeply into the theory and practice of causal discovery. In particular, it has several distinctive features not commonly found elsewhere:

1. Instead of covering all of Bayesian networks, we specialize in the theory and methods of causal discovery. In particular, for LiNGAM and score-based structure learning, we go from theory down to implementation in detail.
2. For kernel-based statistical tests (HSIC, KCI, etc.), we concisely and intuitively describe their mathematical essence with an eye toward applications in the PC algorithm and LiNGAM.

3. As Bayesian approaches to structure learning, we formulate marginal likelihood and its relationship to information criteria (especially BIC), and carefully unpack the underlying theory, including clear implications and limitations of BDeu and Jeffreys priors.
4. We do not stop at introducing concepts; wherever possible we include implementations in R and Python so that readers can learn in a reproducible manner.
5. We keep beginners in mind by explaining key ideas through diagrams and examples to build intuition. Even advanced topics are made more approachable through visual aids and a staged presentation.

While the primary focus is a practical understanding of causal discovery, for readers wishing to dig deeper into theoretical background, the appendices cover advanced theorems and derivations such as:

- The Darmois–Skitovitch theorem
- Murphy's derivation of marginal likelihood (normal mean with inverse-Wishart prior)
- The Verma–Pearl fundamental results on structure learning

These are intended for readers who wish to further enrich their mathematical understanding of causal discovery and are not mandatory for following the main text. Those uneasy with heavy mathematics can safely skip them; those who are curious will find solid, satisfying background explanations.

We summarize the flow of chapters as follows:

Chapter 1 reviews the basics of probability and statistics necessary for understanding the causal discovery methods developed in the second half of the book. Chapter 2 introduces the minimal concepts in graphical models such as conditional independence and separation. Chapter 3 treats kernel-based tests of independence and conditional independence (HSIC, KCI, etc.).

Chapter 4 introduces the PC algorithm that uses these tests. It is a constraint-based structure learning method that identifies the dependency skeleton of variables via conditional independence testing. Chapter 5 covers LiNGAM, which estimates causal order based on non-Gaussianity; in DirectLiNGAM, the independence tests of Chap. 3 play a key role.

Chapter 6 provides the theoretical background for score-based structure learning, covering information criteria (AIC, BIC) and the formulation and properties of marginal likelihood. Chapter 7 presents score-based structure learning methods that leverage these ideas. Like the PC algorithm, they estimate the variable structure, but do so by evaluating scores to compare candidate structures and select an optimal model.

In the figure, red boxes indicate chapters on foundational theory and blue boxes indicate chapters on concrete methods for causal discovery.

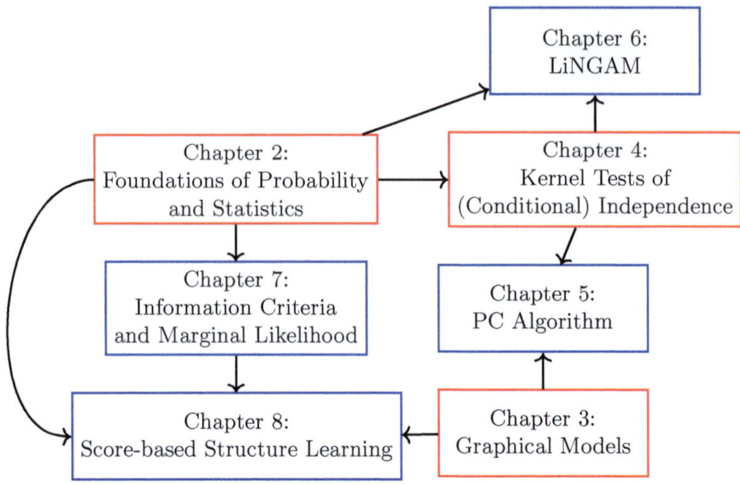

This book aims to face the challenge of causal discovery head-on while conveying to students, practitioners, and researchers alike both the "intellectual elegance" and the "practical power" of graphical models. I sincerely hope that all readers who pick up this volume will deepen their understanding of the rich world of causal discovery.

Features of This Series

Rather than this book alone, the distinctive features of the *series* are summarized below.

1. Make it your own: build the logic
   You grasp concepts mathematically, implement programs, and verify behavior by running them. By repeating this cycle, you will build the "logic" in your mind. You acquire not only knowledge of machine learning but also a way of looking at problems, enabling you to keep pace. Most students say, "I learned a lot," after solving the 100 problems.
2. Not just talk: code so you can move from code to action
   Books on machine learning without source code are inconvenient. Even if a package exists, without source code you cannot improve the algorithm. Sometimes code is on Git, but only for MATLAB or Python, or not sufficient. In this series, most procedures come with code; even if the math is difficult, you can understand what the code is doing.
3. Beyond "how to use": a scholarly text written by a professor
   Books full of package how-tos and examples can help beginners get started, but merely following steps without understanding the underlying operations is unsatisfying. Here, we present the mathematical principles with the code

that realizes them, leaving little room for doubt. This series leans toward the academic, full-fledged end of the spectrum.

4. Solve 100 problems: university exercises refined by student feedback
   The exercises have been used in seminars and lectures, iteratively improved through student feedback, and distilled into an optimal set of 100 problems. Each chapter's main text serves as the explanation; by reading it, you will be able to solve all exercises.
5. Self-contained within the book
   Have you ever felt let down by a proof that says "see Reference ○○ for details"? Unless highly motivated, few readers will track down such references. We choose topics to minimize the need for external citations. Proofs are kept as straightforward derivations; more difficult arguments are placed in appendices at the end of each chapter.
6. Not sell-and-forget: videos, online Q&A, and program files
   In university courses, we answer student questions on Slack around the clock. For this series, we provide a reader page

   https://bayesnet.org/books

   to facilitate casual interaction between authors and readers. We also publish 10–15 minute videos for each chapter. Moreover, programs in the book are available for download from Git.

Toyonaka, Japan Joe Suzuki

Contents

Chapter 1
A Gentle Introduction to Causal Discovery

In this chapter, as a warm-up, I would like to describe the overall picture of causal discovery without going into rigorous details. Using examples, I explain the two problems that make up causal discovery—dependency structure and variable ordering. My goal is for this single chapter to convey not only the big picture of the book but also its essence.

1.1 Causal Discovery = Discovering Dependencies + Discovering Variable Order

By representing the "follow/followed-by" relations on social media as directed edges, we can form a directed network (Fig. 1.1). Such visualizations are meaningful, in that they reveal structures like central "hubs" and "isolated nodes" with no connections, and they can help us discover latent relationships and features.

In contrast, what this book deals with are networks that represent mathematically definable relationships such as probability and causality. In particular, we assume that each vertex corresponds to an actual measured variable (e.g., presence/absence of disease and smoking status). Based on their causal relations (dependencies and ordering), we consider whether to connect each pair of vertices with an edge, and if so, in which direction.

First, a graph that represents dependency structure with undirected edges is called a **Markov network** (MN), and a graph whose edges are all directed and form no cycles (a directed acyclic graph, DAG) is called a **Bayesian network** (BN). Among Bayesian networks, when we orient the arrows from causes to effects, we call it a **causal network**.

Here I will explain the distinction between Bayesian networks (which reflect dependencies) and causal networks (which additionally reflect causation) using the Asia dataset [14]. In Asia, we imagine a physician asking a patient whether

J. Suzuki, *Graphical Models and Causal Discovery with Python*,
https://doi.org/10.1007/978-981-95-5308-2_1

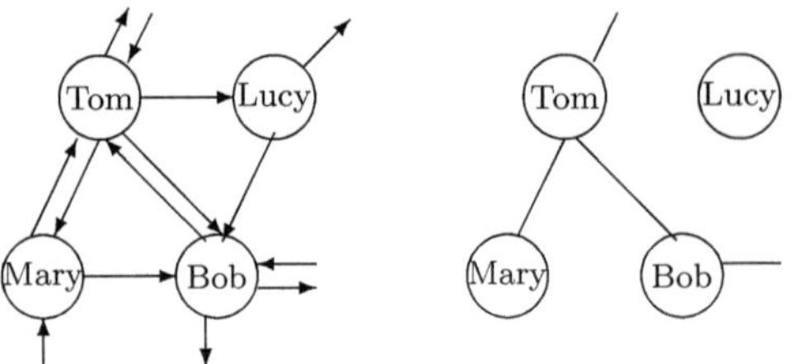

Fig. 1.1 A directed graph representing the relationship on Social Networking Service (SNS) where (following) →(being followed) (left), and an undirected graph showing only the mutual-follow relationships (right). Tom, Lucy, Mary, and Bob are the users

Table 1.1 Variables included in the Asia dataset

Variables	Meaning	Values
`Asia`	Recent travel to Asia	Yes/No
`Smoke`	Smoker	Yes/No
`Tuberculosis`	Tuberculosis	Yes/No
`Lung Cancer`	Lung cancer	Yes/No
`Bronchitis`	Bronchitis	Yes/No
`Either`	Either tuberculosis or lung cancer	Yes/No
`Xray`	Chest X-ray result positive	Positive/Negative
`Dyspnoea`	Shortness of breath	Yes/No

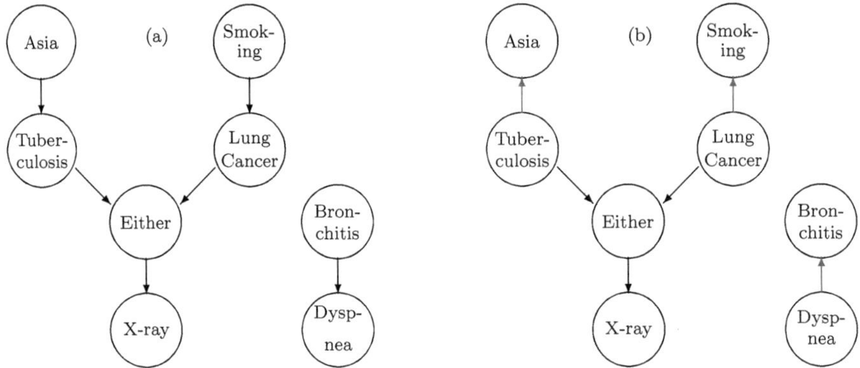

Fig. 1.2 The causal network derived from physicians' beliefs (**a**) and the Markov-equivalent Bayesian network (**b**). In terms of dependencies, as seen from the fact that (1.1) and (1.2) are equal, the two are equivalent Bayesian networks. In the causal network, the arrow directions carry causal meaning, whereas in the Bayesian network this is not necessarily the case

they recently traveled to Asia, whether they smoke, and so on and diagnosing the likelihood of tuberculosis or lung cancer based on that information. Suppose, based on experience, the physician posits the causal relations below among eight random variables as listed in Table 1.1 and constructs the causal network shown in Fig. 1.2a. Throughout, " →" points from the cause on the left to the effect on the right.

Asia → Tuberculosis (Travel to Asia increases the risk of tuberculosis.)
Smoke → Lung Cancer and Smoke → Bronchitis (Smoking increases the risks of lung cancer and bronchitis.)
Tuberculosis → Either and Lung Cancer → Either (If either tuberculosis or lung cancer is present, *Either* becomes Yes.)
Either → Xray (When *Either* is Yes, an X-ray abnormality is more likely.)
Lung Cancer → Dyspnoea and Bronchitis → Dyspnoea (Lung cancer and bronchitis cause shortness of breath.)

The relations in Fig. 1.2 also state that the joint probability of the eight variables (denoted by initial letters) $P(A, S, T, L, B, E, X, D)$ can be written as a product of conditional probabilities:

$$P(X \mid E)P(E \mid T, L)P(T \mid A)P(A)P(L \mid S)P(B)P(S)P(D \mid B). \tag{1.1}$$

The aim of this book is to construct a network like Fig. 1.2a automatically from n i.i.d. (independent and identically distributed) samples on p variables (here, $p = 8$) without relying on domain experts. Sometimes such experts are not available; in other cases, forecasts based on past common sense may no longer apply. In such situations, discovering structure from data is effective.

For example, in structure learning of Bayesian networks studied in Chap. 8, one may conclude that the variable *Asia* is independent of the other seven variables. In the 1980s, travel from Europe or North America to Asia was relatively rare, and such a prior belief might have existed.

However, noting Bayes' theorem,

$$P(T \mid A)P(A) = P(T, A) = P(A \mid T)P(T),$$

we can rewrite (1.1) as

$$P(X \mid E)P(E \mid T, L)P(A \mid T)P(T)P(S \mid L)P(L)P(B \mid D)P(D). \tag{1.2}$$

Thus, Fig. 1.2a and b represents the same dependencies (we say they are **Markov equivalent**).

In general, a Bayesian network expresses dependencies among variables rather than causal order. It expresses a joint distribution as a product of conditionals and depicts that factorization with a DAG. Therefore, for each variable ordering, there is a corresponding Bayesian network.

For instance, consider three variables X, Y, Z connected in that order as a Bayesian network as in Fig. 1.3. In (a),

$$P(X \mid Y)P(Y \mid Z)P(Z) = P(X \mid Y)P(Y)P(Z \mid Y) = P(X)P(Y \mid X)P(Z \mid Y)$$
$$= \frac{P(X, Y)P(Y, Z)}{P(Y)},$$

Fig. 1.3 Bayesian networks in which the random variables X, Y, Z are connected in this order. (**a**), (**b**), and (**c**) represent the same dependency structure. (**d**) is called a collider, in which case X and Z become independent

so it represents the same distribution as (b) and (c). By contrast, the distribution in (d),

$$P(X)P(Y \mid X, Z)P(Z) = \frac{P(X)P(X, Y, Z)P(Z)}{P(X, Z)},$$

differs. At this point, one might object: "Then why do Bayesian networks have arrows at all?" I will clarify this in the latter half of the chapter.

In our running example, Fig. 1.2b is also a valid Bayesian network. From data alone, we can conclude only that the truth lies in one of the Markov-equivalent factorizations like (1.1) or (1.2); in Fig. 1.2b, there are three red directed edges whose directions can be flipped independently, so there are $2^3 = 8$ Markov-equivalent Bayesian networks.

In general, standard statistical knowledge alone cannot deduce variable order. If, however, the causal order of variables is known, then among the Markov-equivalent networks, at most one is compatible with that order, so we need only search for the dependencies. For example, if the order

$$A \to T \to S \to L \to E \to X \to B \to D \tag{1.3}$$

is given, then we select the Bayesian network consistent with this order; it is unique among the eight Markov-equivalent candidates. Nevertheless, having only the order (1.3) is insufficient to recover Fig. 1.2a.

Therefore, to learn a causal network from data, we need the following two stages:

Searching variable order: Under certain assumptions, determine a variable order, and thereby identify a single member among the Markov-equivalent Bayesian networks. (Chap. 6)

Searching dependencies: Determine the dependencies among variables and narrow down to the (multiple) Markov-equivalent Bayesian networks (Chaps. 5 and 8).

1.2 Searching Variable Order

Given data, standard statistical methods cannot identify the variable order. Since

$$P(X, Y) = P(X)P(Y \mid X) = P(X \mid Y)P(Y),$$

we cannot tell which factorization is appropriate. In this book, we assume the following for order discovery. Let X, Y be zero-mean continuous random variables. We posit there exist a constant $a \neq 0$ and a random variable e independent of X such that

$$Y = aX + e. \tag{1.4}$$

When this holds, X is the cause and Y is the effect. By symmetry, swapping X and Y gives the following: There exist a constant $a' \neq 0$ and a random variable e' independent of Y such that

$$X = a'Y + e'. \tag{1.5}$$

When this holds, Y is the cause and X is the effect. This is the *additive noise model*. It is the criterion underlying LiNGAM discussed in this book, originally proposed by H. Kano in 2002.

You may wonder whether (1.4) and (1.5) could both hold simultaneously. The answer is clear:

Claim 1

(X, Y) are Gaussian $\Longleftrightarrow$ (1.4) and (1.5) both hold.

We explain why in Chap. 6. Furthermore, assume at least one of (1.4) or (1.5) holds. Then Claim 1 implies:

Claim 2
If at least one of (1.4) or (1.5) holds, then

At least one of X, Y is non-Gaussian

$\Longleftrightarrow$ The causal direction between X and Y is identifiable.

Thus, as long as we avoid the case where *both* X and Y are Gaussian, the cause–effect direction is identifiable; non-identifiability occurs only when both are Gaussian.

Based on this fact, proceed as follows. Given samples $(X, Y) = (x_1, y_1), \ldots, (x_n, y_n)$, let $x^n := (x_1, \ldots, x_n)$ and $y^n := (y_1, \ldots, y_n)$. Define the regression coefficients a, a' via

$$c(x^n, y^n) := \frac{1}{n}\sum_{i=1}^{n} x_i y_i, \quad v(x^n) := \frac{1}{n}\sum_{i=1}^{n} x_i^2, \quad v(y^n) := \frac{1}{n}\sum_{i=1}^{n} y_i^2, \tag{1.6}$$

and set

$$a := \frac{c(x^n, y^n)}{v(x^n)}, \qquad a' := \frac{c(x^n, y^n)}{v(y^n)}.$$

Compute residual vectors (each of length n)

$$e := y^n - ax^n, \qquad e' := x^n - a'y^n.$$

Since we assume zero means, we do not subtract sample means $\overline{x} := n^{-1}\sum_i x_i$ and $\overline{y} := n^{-1}\sum_i y_i$ in (1.6).

We then test which pair is independent, (e, x^n) or (e', y^n)—that is, whether (1.4) or (1.5) holds. By Claim 2, we can determine which is the cause and which is the effect, provided the sample size n is sufficiently large.

This requires a test of independence. If $X \perp\!\!\!\perp Y$, then their correlation is zero, but not conversely. In our construction, both $c(e, x^n)$ and $c(e', y^n)$ equal zero; for example,

$$c(e, x^n) = \frac{1}{n}\sum_{i=1}^{n} e_i x_i = \frac{1}{n}\sum_{i=1}^{n}(y_i - ax_i)x_i = c(x^n, y^n) - a\, v(x^n) = 0.$$

Similarly $c(e', y^n) = 0$. When (X, Y) are Gaussian, independence is equivalent to zero correlation; this aligns with Claim 2.

Therefore, independence testing based on (near-)zero correlation is inadequate: We would fail to detect dependence whenever correlation happens to be zero. Such tests are meaningful only under joint Gaussianity and are inappropriate here.

A widely used alternative is **HSIC** (Hilbert–Schmidt Independence Criterion). Let $k : \mathcal{X} \times \mathcal{X} \to \mathbb{R}$ be symmetric ($k(x, y) = k(y, x)$) and positive-definite (for any $n \geq 1$ and $x_1, \ldots, x_n \in \mathcal{X}$, the matrix $(k(x_i, x_j))$ is positive semidefinite). We call such k a (positive-definite) kernel. To assess independence of random variables $X \in \mathcal{X}$ and $Y \in \mathcal{Y}$, we prepare kernels $k_{\mathcal{X}}$ and $k_{\mathcal{Y}}$ and examine the independence of the feature functions $k_{\mathcal{X}}(X, \cdot)$ and $k_{\mathcal{Y}}(Y, \cdot)$. HSIC captures nonlinear relations and has strong power in this setting. We study it in detail in Chap. 4.

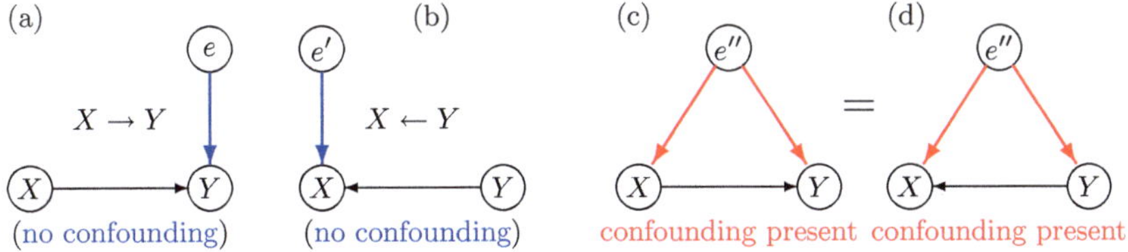

Fig. 1.4 Panel (**a**) represents, by a Bayesian network, the dependency when there exist constants a, a' such that $Y = aX + e$ with $X \perp\!\!\!\perp e$, and panel (**b**) represents the case $X = a'Y + e'$ with $Y \perp\!\!\!\perp e'$. In each case, there is a collider at Y or X, respectively. In this situation, noise is added only to either X or Y (i.e., there is no confounding). On the other hand, panels (**c**) and (**d**) show the case where noise is added to both X and Y (i.e., there is confounding). Either of the Markov-equivalent Bayesian networks (**c**) or (**d**) can be used to represent this

This procedure is known as Direct LiNGAM. **LiNGAM** (Linear Non-Gaussian Acyclic Model) was proposed by Shimizu et al. in 2006 and has several variants. The original approach, ICA-LiNGAM, is based on independent component analysis. We introduce both but focus on Direct LiNGAM for its intuitive clarity.

The dependency patterns (1.4) and (1.5) correspond to Fig. 1.4a and b, where noise enters only one of X or Y. When neither (1.4) nor (1.5) holds, the situation is like Fig. 1.4c and d: Noise affects both X and Y. We call this **confounding**. Standard LiNGAM assumes no confounding; we also discuss how to handle confounding in this book.

For multiple variables, consider X, Y, Z. There are six possible orders. If $X \to Y \to Z$, then

$$X = e_X, \qquad Y = aX + e_Y, \qquad Z = bX + cY + e_Z,$$

and we ask whether there exist constants a, b, c such that e_X, e_Y, e_Z are mutually independent.

Proceed from the upstream variable. Given samples $x^n = (x_1, \ldots, x_n)$, $y^n = (y_1, \ldots, y_n)$, $z^n = (z_1, \ldots, z_n)$, define

$$b := \frac{c(x^n, y^n)}{v(x^n)}, \qquad c := \frac{c(x^n, z^n)}{v(x^n)},$$

and compute residuals

$$y^n_x := y^n - bx^n, \qquad z^n_x := z^n - cx^n.$$

Test whether x^n is independent of $\{y^n_x, z^n_x\}$ (the latter can be viewed as an $n \times 2$ matrix). Do likewise for Y and Z: test y^n vs. $\{z^n_y, x^n_y\}$ and z^n vs. $\{x^n_z, y^n_z\}$. For convenience, you may compute HSIC for each of the three pairs and choose the one with the smallest value.

Suppose X is found to be the most upstream. Then set

$$b := \frac{c(y_x^n, z_x^n)}{v(z_x^n)}, \qquad c := \frac{c(y_x^n, z_x^n)}{v(y_x^n)},$$

and residuals

$$y_{xz}^n := y_x^n - bz_x^n, \qquad z_{xy}^n := z_x^n - cy_x^n.$$

Test independence between y_x^n and z_{xy}^n and between z_x^n and y_{xz}^n. If, say, the former are independent (so Y is upstream of Z), then the overall order is $X \to Y \to Z$.

In this way, for any number of variables, we can determine the order from residual-based independence tests while removing the influence of already-selected upstream variables.

1.3 Searching Dependencies

Once the variable order is fixed, we only need to determine, for each variable, which upstream variables it depends on (its parent set).

In the Asia dataset, given the order (1.3), A has an empty parent set; T has either $\{A\}$ or $\emptyset$; S has one of $\{A, T\}$, $\{A\}$, $\{T\}$, or $\emptyset$; and so on. For each variable, we select a subset of the upstream variables.

We then score candidate parent sets using **information criteria** (AIC or BIC) or **marginal likelihood** and choose the best parent set.

A brief detour: Very few people doubt "Newton's laws of motion." Newtonian mechanics has explained real-world motion with remarkable accuracy for centuries. One reason such laws are widely accepted is their *parsimony*: A few simple laws explain many phenomena.

Model selection in statistics follows a similar philosophy: In addition to goodness of fit, we value how simple the model is. Concretely, for a model M, we evaluate lack of fit by the negative log-likelihood H_M and complexity by the number of parameters d_M (smaller sums are better). Information criteria and marginal likelihood balance these two to choose a model with high explanatory power yet "just-right" complexity.

AIC (Akaike's Information Criterion) and BIC (Bayesian Information Criterion) weight H_M and d_M with ratios 1:1 and $1{:}\frac{1}{2}\log n$, respectively, where n is the sample size:

$$AIC_M = H_M + d_M, \qquad BIC_M = H_M + \frac{d_M}{2}\log n. \tag{1.7}$$

We then select the model M minimizing each score. Suppose, for variable S, the parent set M induces m conditioning states (e.g., with parents A, T, there are four combinations). In Asia, the candidates for the parent set of S are $\{A, T\}$, $\{A\}$, $\{T\}$, and $\{\}$, with $m = 4, 2, 2, 1$, respectively.

If the parent set is empty, let θ be $P(S = 1)$ and c its count in the data. The likelihood is

$$\prod_{i=1}^{n} P(x_i \mid \theta) = \theta^c (1-\theta)^{n-c}.$$

Differentiating the log-likelihood w.r.t. θ and setting to zero yield the Maximum Likelihood Estimate (MLE)

$$\frac{\partial}{\partial \theta} \log\{\theta^c (1-\theta)^{n-c}\} = \frac{c}{\theta} - \frac{n-c}{1-\theta} = 0 \Rightarrow \theta = \frac{c}{n}.$$

The negative log-likelihood is

$$\mathcal{H}_2(p) := -p \log p - (1-p)\log(1-p),$$

so $H_M = n\mathcal{H}_2(c/n)$. For $m \geq 2$, let n_j be the frequency of state j and c_j the frequency of $S = 1$ within that state; then $H_M = \sum_{j=1}^{m} n_j \mathcal{H}_2(c_j/n_j)$. From the dataset we obtain the following counts:

A	T	$S=0$	$S=1$
0	0	2448	2468
0	1	21	21
1	0	15	25
1	1	1	1

A	$S=0$	$S=1$
0	2469	2489
1	16	26

T	$S=0$	$S=1$
0	2463	2493
1	22	22

$S=0$	$S=1$
2485	2515

Since each variable is binary and each state contributes one probability parameter, when the parent set of M has m states, $d_M = m$. The resulting AIC and BIC are as follows:

M	H_M	d_M	AIC_M	BIC_M
$\{A, T\}$	4997.69	4	5001.69	5014.72
$\{A\}$	4997.96	2	4999.96	5006.48
$\{T\}$	4999.51	2	5001.51	5008.03
$\{\}$	4999.50	1	5000.50	5003.76

Thus, with AIC we draw a directed edge $A \to S$, while with BIC we draw no incoming edge to S. Although Asia is entirely binary, H_M and d_M are defined similarly for general multinomial variables (see Chap. 8).

Given the order, performing this parent-selection step for every variable yields a Bayesian network.

For continuous variables, proceed similarly. Let $(X_j)_{j\in\pi}$ be the parent set of X_k. From the samples $x_{i,j}$ $(j \in \pi)$ and $x_{i,k}$ $(i = 1, \dots, n)$,

$$\hat{\sigma}^2 = \min_{\beta_j,\, j\in\pi} \left\{ \sum_{i=1}^{n} \Big(x_{i,k} - \sum_{j\in\pi} x_{i,j} \Big)^2 \right\},$$

and set $H_M = 0.5 \log \hat{\sigma}^2$ and $d_M = |\pi|$.

Some references define AIC as minimizing $2H_M + 2d_M$ or equivalently maximizing $-2H_M - 2d_M$; the optimizer M is the same. We use the notation in (1.7) throughout.

We also consider dependency search via marginal likelihood. For i.i.d. length-3 binary sequences (eight patterns), assign

$$P(x_1x_2x_3 \mid \theta) = \theta^k(1-\theta)^{3-k}, \quad 0 \le k \le 3,$$

and place the prior

$$\frac{1}{\pi\sqrt{\theta(1-\theta)}}, \qquad 0 \le \theta \le 1.$$

Then

$$\int_0^1 \theta^k(1-\theta)^{3-k} \cdot \frac{1}{\pi\sqrt{\theta(1-\theta)}}\, d\theta = \int_0^1 \theta^{k-0.5}(1-\theta)^{2.5-k}\, d\theta$$

evaluates to $3/8$ for $k = 0, 3$ and $1/8$ for $k = 1, 2$. This is the marginal likelihood. With a chosen prior for parameters, we can compute it for any length n sequence x^n and likewise for paired data (x_i, y_i). Denote $Q(x^n, y^n)$ as the marginal likelihood for (X, Y); the definition extends to discrete and continuous variables.

Using marginal likelihoods, with prior probability $0 < r < 1$ for independence, one can show that, with probability 1 as $n \to \infty$,

$$r\, Q(x^n)Q(y^n) \le (1-r)\, Q(x^n, y^n)$$

if and only if X and Y are independent. The asymptotic performance does not depend on r as long as $r \notin \{0, 1\}$.

Moreover, when selecting the parent set of Z from subsets of $\{X, Y\}$, compare

$$\frac{Q(x^n, y^n, z^n)}{Q(x^n, y^n)}, \qquad \frac{Q(x^n, z^n)}{Q(x^n)}, \qquad \frac{Q(y^n, z^n)}{Q(y^n)}, \qquad Q(z^n),$$

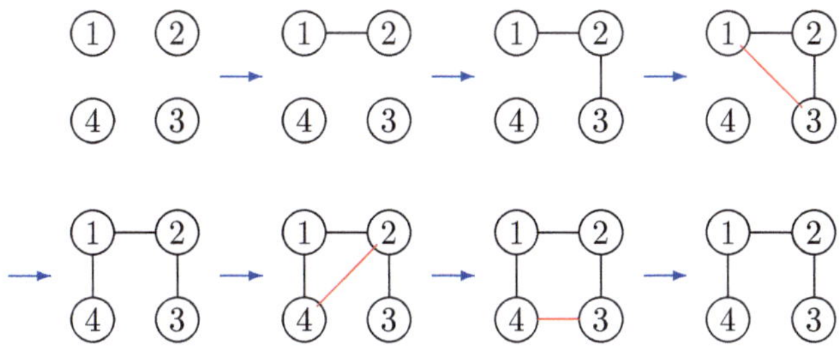

Fig. 1.5 For the random variables X_1, X_2, X_3, X_4, suppose the mutual information satisfies $I(X_1, X_2) > I(X_2, X_3) > I(X_1, X_3) > I(X_1, X_4) > I(X_2, X_4) > I(X_3, X_4)$. Edges $\{1, 2\}$, $\{2, 3\}$, $\{1, 3\}$, $\{1, 4\}$, $\{2, 4\}$, and $\{3, 4\}$ are added in this order. When attempting to add $\{1, 3\}$, a cycle is created, so $\{1, 3\}$ is not included. The same applies to $\{2, 4\}$ and $\{3, 4\}$

and choose among $\{X, Y\}$, $\{X\}$, $\{Y\}$, or $\emptyset$ according to which is largest. Crucially, one can prove that such parent-selection converges to the true dependency as $n \to \infty$. BIC approximates the negative log of the marginal likelihood and thus has similar asymptotic behavior.

A drawback of estimating the exact BN structure by information criteria or marginal likelihood is the enormous computation time—exponential in p (Fig. 1.6). One can restrict the search space (e.g., with hill climbing) while scoring by an information criterion.

Another approach is the **Chow–Liu algorithm**, which constructs a tree from the pairwise mutual information $I(X_i, X_j)$ for $1 \le i < j \le p$. Add undirected edges in order of decreasing $I(X_i, X_j)$, skipping any edge that would create a cycle (Fig. 1.5). The resulting tree maximizes the sum of edge mutual information and, moreover, is guaranteed to minimize the Kullback–Leibler divergence from the true distribution among all trees (Chap. 3).

From marginal likelihoods, we can also define an estimator of mutual information

$$I(X, Y) = \sum_k \sum_l P(X = k, Y = l) \log \frac{P(X = k, Y = l)}{P(X = k) P(Y = l)}.$$

Given counts $n_{k,l}$ for $(X, Y) = (k, l)$ (with $n_{k,\cdot} = \sum_l n_{k,l}$ and $n_{\cdot,l} = \sum_k n_{k,l}$), the MLE-based estimator

$$I_n := \sum_k \sum_l \frac{n_{k,l}}{n} \log \frac{(n_{k,l}/n)}{(n_{k,\cdot}/n)\,(n_{\cdot,l}/n)}$$

is commonly used; indeed $I_n \to I(X, Y)$ (consistency). Alternatively, using marginal likelihoods, define

$$J_n := \frac{1}{n} \log \frac{Q(x^n, y^n)}{Q(x^n) Q(y^n)} \approx I_n - \frac{(\alpha - 1)(\beta - 1)}{2n} \log n,$$

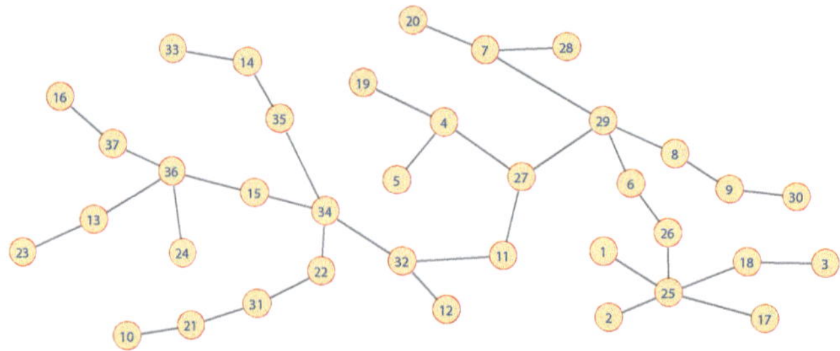

Fig. 1.6 Alarm network ($p = 37$). Computing the exact optimum by information criteria or marginal likelihood takes over 24 hours already at $p = 30$, about 2 days at $p = 31$, and 4 days at $p = 32$—exponential in p

where α, β are the numbers of categories of X and Y. This yields not only consistency but also *detection* of independence: With probability 1,

$$X \perp\!\!\!\perp Y \iff J_n \leq 0.$$

Thus, if $J_n \leq 0$, we do not connect the pair—generalizing Chow–Liu to produce a forest rather than a tree. When $X \perp\!\!\!\perp Y$, $I_n \to 0$ but remains positive, so I_n alone cannot decide independence. For continuous variables, an analogous method applies; set

$$J_n = I_n - \frac{1}{2n} \log n.$$

This forest-construction algorithm scales roughly with p^2 rather than exponentially in p (Fig. 1.6).

1.4 Graphical Models and Conditional Independence

At the outset I mentioned that arrows in a Bayesian network do not necessarily represent causal directions. Let me clarify this point to close the chapter.

Bayesian networks admit an equivalent definition: They are DAGs that jointly encode the conditional independences among $X_1, \ldots, X_p$. Markov networks likewise encode conditional independences with an undirected graph.

Whereas independence may be intuitive, conditional independence can be harder to grasp. Events A and B are independent if $P(A, B) = P(A)P(B)$. Events A and C are conditionally independent given B if

$$P(A, B, C) = \frac{P(A, B)P(B, C)}{P(B)}. \tag{1.8}$$

Fig. 1.7 Representation of separations using DAGs. (**a**), (**b**), and (**c**) represent $1 \perp\!\!\!\perp 3 \mid 2$, while (**d**) represents $1 \perp\!\!\!\perp 3$

As an example, let A, B, and C denote rain on January 1, 2, and 3. Since weather evolves continuously, the weather on the 3rd may be predictable from that on the 2nd without knowing the 1st. We write this as $P(C \mid B) = P(C \mid A, B)$ or (1.8). Conditional independence is defined similarly for (sets of) random variables, and we can represent it graphically.

First consider an undirected graph G. For disjoint index sets $A, B, C \subset \{1, \ldots, p\}$, if every path from any $i \in A$ to any $j \in C$ passes through some vertex in B, then we say A and C are separated by B and write $A \perp\!\!\!\perp_G C \mid B$. Let $X_A := \{X_i\}_{i \in A}$, $X_B := \{X_j\}_{j \in B}$, $X_C := \{X_k\}_{k \in C}$. We aim to have

$$A \perp\!\!\!\perp_G C \mid B \Longrightarrow X_A \perp\!\!\!\perp X_C \mid X_B \tag{1.9}$$

hold as much as possible for the distribution of $X_1, \ldots, X_p$. If all pairs are connected, there is no separation, so (1.9) holds vacuously. As we remove edges, (1.9) may fail. The Markov network is the sparsest undirected graph G for which (1.9) still holds.

For Bayesian networks, a similar condition holds, but separation is defined somewhat differently for DAGs. In Fig. 1.7a,b,c, separation is as in undirected graphs; in the collider case (d), however, separation flips depending on whether node ② or its descendants are conditioned on. The precise definition is given in Chap. 3; this notion is called **d-separation**. Using it, we obtain the conditional independences encoded by a BN. For large DAGs, reading $A \perp\!\!\!\perp_G C \mid B$ off the graph may be nontrivial (not mathematically difficult, but it requires practice).

Note that a single undirected graph or DAG typically encodes multiple conditional independences. Also, the sets of independences representable by Markov vs. Bayesian networks differ; an implication like (1.9) with "$\Longleftrightarrow$" may hold for a BN but not for a Markov network and vice versa.

So far we have explored dependency discovery by optimizing a score such as an information criterion or marginal likelihood (the *score-based* approach). An alternative is the *constraint-based* approach, which constructs a BN by testing conditional independences from data.

A well-known constraint-based method is the **PC algorithm** (Peter–Clark). It assumes (for the underlying distribution) that the "$\Longleftrightarrow$" in (1.9) holds (the *faithfulness* assumption). In brief (details in Chap. 5), we perform tests of the form

$$X_i \perp\!\!\!\perp X_j \mid (X_k)_{k \in \pi}, \qquad \pi \subseteq \{1, \ldots, p\} \setminus \{i, j\}, \tag{1.10}$$

increasing $|\pi|$ up to the largest set for which (1.10) still holds. Under faithfulness, all conditional independences are of this form (Chap. 5). After these tests, for each $i \neq j$ we obtain a maximal separating set $\pi(i, j)$ with $X_i \perp\!\!\!\perp X_j \mid \pi(i, j)$. However, this yields only a Markov equivalence class of BNs; without variable-order discovery, causal discovery remains incomplete.

As with independence testing, conditional independence cannot be reliably inferred from small correlations alone. We therefore apply **KCI** (Kernel-Based Conditional Independence test), an extension of HSIC (Chap. 4).

In this chapter we presented an intuitive overview of causal discovery. From the next chapter, we return to the basics of probability and statistics and formally develop the prerequisites for causal discovery.

The causal discovery treated in this book deals with graphical models and differs from the causal inference in the sense of Donald Rubin. It is often referred to as Pearl-style causal inference [19].

Chapter 2
Foundations of Probability and Statistics

In this chapter, as the groundwork for learning causal discovery with graphical models, we introduce the basic concepts of probability and statistics. We begin by defining fundamental terms such as events, probabilities, and random variables and then introduce representative probability distributions: the binomial, normal, Poisson, and Gamma distributions. We also discuss concepts used to describe relations among multiple random variables—joint distributions, independence, correlation coefficients, and covariance matrices.

We further present properties of the normal distribution—indispensable in multivariate analysis and causal inference—and its behavior under linear transformations, thereby laying the mathematical foundation for describing probabilistic structure. We briefly touch on the Wishart distribution and its inverse, showing their connection to Bayesian statistics.

In the second half, we introduce maximum likelihood estimation as a method to estimate model parameters from observed data and explain the framework of statistical testing. We define Type I/II errors, significance level, and power and provide an overall understanding of hypothesis testing.

Below, we denote by $\mathbb{R}$, $\mathbb{Q}$, $\mathbb{Z}$, and $\mathbb{N}$ the sets of all real numbers, rational numbers, integers, and nonnegative integers, respectively. For positive integers l, m, n, we write $\mathbb{R}^{l\times m}$ and $\mathbb{R}^n$ for the set of real $l \times m$ matrices and the set of (column) vectors of length n. For $a < b$, the intervals (a, b), $(a, b]$, $[a, b)$, and $[a, b]$ denote $\{x \in \mathbb{R} : a < x < b\}$, $\{x \in \mathbb{R} : a < x \leq b\}$, $\{x \in \mathbb{R} : a \leq x < b\}$, and $\{x \in \mathbb{R} : a \leq x \leq b\}$, respectively.

2.1 Random Variables

When rolling a die, one of the faces $1, 2, 3, 4, 5, 6$ occurs at random. A specific condition regarding the outcome (e.g., "an even face occurs" or "a 1 or 2 occurs")

J. Suzuki, *Graphical Models and Causal Discovery with Python*,
https://doi.org/10.1007/978-981-95-5308-2_2

is called an **event**. We call the entire set, such as $\{1, 2, 3, 4, 5, 6\}$, the **sample space** and denote it by Ω. We represent events as subsets like $\{2, 4, 6\}$ or $\{1, 2\}$ (there are 2^6 such subsets). The family $\mathcal{F}$ whose elements are these events is called the σ-algebra (collection of events). Thus

$$\Omega = \{1, 2, 3, 4, 5, 6\}, \qquad \mathcal{F} = \underbrace{\{\{\}, \{1\}, \{2\}, \{3\}, \{4\}, \{5\}, \{6\}, \ldots, \{1, 2, 3, 4, 5, 6\}\}}_{2^6 \text{ elements}},$$

where $\{\}$ denotes the empty set. Each event $A \in \mathcal{F}$ is a subset of Ω, and $\mathcal{F}$ must be closed under intersection ($\cap$), union ($\cup$), and complementation ($\bar{\ }$): namely,

$$A \in \mathcal{F} \Longrightarrow \bar{A} \in \mathcal{F}, \qquad A, B \in \mathcal{F} \Longrightarrow \begin{cases} A \cap B \in \mathcal{F} \\ A \cup B \in \mathcal{F} \end{cases}$$

A map $P : \mathcal{F} \to [0, 1]$ is called a **probability** if it assigns to each event $A \in \mathcal{F}$ a value $P(A)$ satisfying:

1. $P(A) \geq 0$ for all $A \in \mathcal{F}$.
2. $P(\Omega) = 1$.
3. For pairwise disjoint $A_1, A_2, \ldots$,

$$P\left(\bigcup_{i=1}^{\infty} A_i\right) = \sum_{i=1}^{\infty} P(A_i).$$

If the die is biased and we estimate probabilities empirically, we must assign 2^6 probabilities satisfying the three axioms above.

From these axioms it follows that

$$P(\{\}) = 0 \tag{2.1}$$

and, for all $A, B \in \mathcal{F}$,

$$P(A \cup B) = P(A) + P(B) - P(A \cap B) \tag{2.2}$$

(Problem 1).

In what follows we assume the triple $(\Omega, \mathcal{F}, P)$ is given.[1]

Return to the die. Define a variable X that takes $X = 0$ for even outcomes and $X = 1$ for odd outcomes:

$$\{1, 2, 3, 4, 5, 6\} \ni \omega \mapsto X(\omega) = \begin{cases} 0, & \omega = 2, 4, 6 \\ 1, & \omega = 1, 3, 5 \end{cases}$$

[1] Called a **probability space**.

A map $X : \Omega \to \mathbb{R}$ of this kind is called a **random variable** and is used to define events; for example, $X = 1$ corresponds to the event $\{1, 3, 5\}$. We write random variables in uppercase and their realizations in lowercase.

We distinguish **finite** vs. **infinite** sets, and among infinite sets, **countable** vs. **uncountable**. A countable set can be put in one-to-one correspondence with $\mathbb{N}$. $\mathbb{Q}$ is countable, and $\mathbb{R}$ is uncountable. Finite or countable sets are said to have **at most countably many** elements.

Without loss of generality, we assume random variables take real values. A random variable taking values in a finite set (identified with nonnegative integers up to some bound) or in a countable set (identified with nonnegative integers) is called a **discrete random variable**. In the die example, X is discrete with values $\{0, 1\}$. In Examples 2.1 and 2.2, X takes finitely and infinitely many values, respectively. We may assume X takes values $0, 1, 2, \ldots$, and if it takes only m values, set $p_m = p_{m+1} = \cdots = 0$. From the axioms, $p_k \geq 0$ and

$$\sum_{k=0}^{\infty} p_k = 1 \tag{2.3}$$

are required.

Example 2.1 (Binomial Distribution) Suppose a wrestler wins a bout with probability $0 < p < 1$ independently of the opponent. After $n(\geq 1)$ bouts, the probability of k wins and $n - k$ losses $(0 \leq k \leq n)$ is

$$p_k = \frac{n!}{k!(n-k)!} p^k (1-p)^{n-k}.$$

Indeed, by the binomial theorem,

$$1 = (p + 1 - p)^n = \sum_{k=0}^{n} \frac{n!}{k!(n-k)!} p^k (1-p)^{n-k} = \sum_{k=0}^{n} p_k,$$

so (2.3) holds. ■

Example 2.2 (Poisson Distribution) Let $X = 0, 1, 2, \ldots$ denote the number of customers waiting at a store. Often we assume for some $\lambda > 0$

$$p_k = e^{-\lambda} \frac{\lambda^k}{k!}.$$

Using the Maclaurin expansion $e^x = \sum_{k=0}^{\infty} x^k / k!$ with $x = \lambda$ yields (2.3). ■

When, for all $x \in \mathbb{R}$, the cumulative probability $P(X \leq x)$ can be written with an integrable function f_X over its domain as

$$P(X \leq x) = \int_{-\infty}^{x} f_X(t)\, dt, \tag{2.4}$$

f_X is called the **probability density function** (pdf) of X, and X is a **continuous random variable**. From the axioms, $f_X(x) \geq 0$ and

$$1 = \lim_{x \to \infty} P(X \leq x) = \int_{-\infty}^{\infty} f_X(x)\,dx \tag{2.5}$$

must hold.

Change of variables is often needed. In two dimensions, for $g, h : \mathbb{R}^2 \to \mathbb{R}$,

$$\int_{x_1}^{x_2} \int_{y_1}^{y_2} f(x, y)\,dx\,dy = \int_{u_1}^{u_2} \int_{v_1}^{v_2} f\big(g(u, v), h(u, v)\big)\,|J|\,du\,dv,$$

where

$$x_1 = g(u_1, v_1),\ x_2 = g(u_2, v_2),\ y_1 = h(u_1, v_1),\ y_2 = h(u_2, v_2),$$

and J is the **Jacobian**

$$J = \det \begin{bmatrix} \dfrac{\partial g(u, v)}{\partial u} & \dfrac{\partial g(u, v)}{\partial v} \\ \dfrac{\partial h(u, v)}{\partial u} & \dfrac{\partial h(u, v)}{\partial v} \end{bmatrix} = \frac{\partial g}{\partial u} \cdot \frac{\partial h}{\partial v} - \frac{\partial g}{\partial v} \cdot \frac{\partial h}{\partial u}.$$

Example 2.3 (Standard Normal Distribution) Let

$$f_X(x) = \frac{1}{\sqrt{2\pi}} \exp\left(-\frac{x^2}{2}\right)$$

be the pdf of continuous X; we say X follows the **standard normal distribution**. For $r > 0$, $0 \leq \theta \leq \pi/2$, set $x = r\cos\theta$, $y = r\sin\theta$. The Jacobian is

$$J = \det \begin{bmatrix} \dfrac{\partial x}{\partial r} & \dfrac{\partial x}{\partial \theta} \\ \dfrac{\partial y}{\partial r} & \dfrac{\partial y}{\partial \theta} \end{bmatrix} = \det \begin{bmatrix} \cos\theta & -r\sin\theta \\ \sin\theta & r\cos\theta \end{bmatrix} = r > 0.$$

Since $x^2 + y^2 = r^2$,

$$\begin{aligned} \left\{\int_{-\infty}^{\infty} \exp\left(-\frac{x^2}{2}\right) dx\right\}^2 &= \int_{-\infty}^{\infty} \int_{-\infty}^{\infty} \exp\left(-\frac{x^2 + y^2}{2}\right) dx\,dy \\ &= 4\int_0^{\infty} \int_0^{\pi/2} \exp\left(-\frac{r^2}{2}\right) r\,dr\,d\theta = 2\pi \end{aligned} \tag{2.6}$$

(Problem 2), which gives (2.5). ■

Next, for $\alpha > 0$, define the **Gamma function**

$$\Gamma(\alpha) := \int_0^\infty t^{\alpha-1} e^{-t}\, dt. \tag{2.7}$$

For $\alpha > 0$,

$$\Gamma(\alpha + 1) = \alpha\, \Gamma(\alpha). \tag{2.8}$$

In particular, for nonnegative integers n, $\Gamma(n+1) = n!$ (Problem 3); also $\Gamma(1/2) = \sqrt{\pi}$ (Problem 3). If

$$f_X(x) = \frac{1}{\Gamma(a)} b^a x^{a-1} e^{-bx}, \quad x > 0, \tag{2.9}$$

then X has a **Gamma distribution**, written $X \sim G(a, b)$. Setting $a = n/2$ and $b = 1/2$ yields the **chi-square distribution** with n degrees of freedom, $X \sim \chi^2_n$, having density

$$f_X(x) = \frac{1}{2^{n/2}\Gamma(n/2)} x^{n/2-1} e^{-x/2}, \quad x > 0. \tag{2.10}$$

For $a, b > 0$, define the **Beta function**

$$B(a, b) := \int_0^1 \theta^{a-1}(1-\theta)^{b-1}\, d\theta.$$

Without proof, it relates to Γ as

$$B(a, b) = \frac{\Gamma(a)\Gamma(b)}{\Gamma(a+b)}. \tag{2.11}$$

The **Beta distribution** with density

$$f_X(x) = \frac{x^{a-1}(1-x)^{b-1}}{B(a, b)}, \quad 0 < x < 1$$

is written $X \sim Be(a, b)$.

Not all random variables are discrete or continuous.

Example 2.4 Let $X = -1$ with probability $1/2$, and $0 \le X \le 1$ with probability $1/2$ having density $f_X(x) = 1/2$ on $[0, 1]$. Then

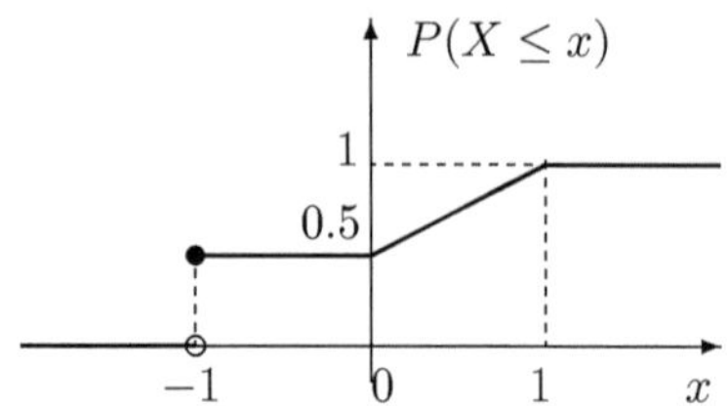

Fig. 2.1 For the random variable X in Example 2.5, there does not exist a probability density function f_X: $P(X \leq x)$ which is discontinuous at $x = -1$

$$P(X \leq x) = \begin{cases} 0, & x < -1 \\ \dfrac{1}{2}, & -1 \leq x < 0 \\ \dfrac{1+x}{2}, & 0 \leq x < 1 \\ 1, & 1 \leq x \end{cases}$$

X takes uncountably many values, but no f_X exists satisfying (2.4) (Fig. 2.1); thus X is not continuous. ■

Random variables that are neither purely discrete nor continuous are inconvenient; hence, when multiple variables appear, we handle either all-discrete or all-continuous cases.

A continuous function of a random variable is also a random variable.[2] For example, scalar multiples, shifts, and squares of X are random variables.

Define the **expectation** $\mathbb{E}[X]$ of X[3] by

$$\mathbb{E}[X] := \begin{cases} \displaystyle\sum_{k=0}^{\infty} k\, P(X = k), \ X \text{ discrete} \\ \displaystyle\int_{-\infty}^{\infty} x f_X(x)\, dx, \ X \text{ continuous.} \end{cases}$$

Let $\mu = \mathbb{E}[X]$. Then $(X - \mu)^2$ is a random variable, and its expectation $\mathbb{V}[X] = \mathbb{E}[(X - \mu)^2]$ is the **variance** of X:

$$\mathbb{V}[X] := \begin{cases} \displaystyle\sum_{k=0}^{\infty} (k - \mu)^2 P(X = k), \ \ X \text{ discrete} \\ \displaystyle\int_{-\infty}^{\infty} (x - \mu)^2 f_X(x)\, dx, \ X \text{ continuous.} \end{cases}$$

[2] In general, measurability is required so that the induced events belong to the predetermined σ-algebra $\mathcal{F}$ (measurability). It is common to assume that $\mathcal{F}$ is closed under (at most) countable unions.

[3] For continuous X, defined when $\int_{-\infty}^{\infty} |x| f_X(x)\, dx < \infty$.

For $a, b \in \mathbb{R}$,

$$\mathbb{E}[aX + b] = a\,\mathbb{E}[X] + b \tag{2.12}$$

$$\mathbb{V}[aX + b] = a^2\,\mathbb{V}[X] \tag{2.13}$$

$$\mathbb{V}[X] = \mathbb{E}[X^2] - \{\mathbb{E}[X]\}^2 \tag{2.14}$$

(Problem 4).

Example 2.5 (Normal Distribution) For standard normal X, $f_X(x)$ is even and $xf_X(x)$ is odd, so $\mathbb{E}[X] = 0$. Using (2.5),

$$\int_{-\infty}^{\infty} x^2 f_X(x)\,dx = \int_{-\infty}^{\infty} x\cdot\{-f_X(x)\}'\,dx = [x\cdot\{-f_X(x)\}]_{-\infty}^{\infty} + \int_{-\infty}^{\infty} f_X(x)\,dx = 1,$$

so $\mathbb{V}[X] = 1$. For $\sigma > 0$, $\mu \in \mathbb{R}$, with $y = \sigma x + \mu$ and $t = \sigma s + \mu$,

$$\int_{-\infty}^{x} \frac{1}{\sqrt{2\pi}} e^{-s^2/2}\,ds = \int_{-\infty}^{y} \frac{1}{\sqrt{2\pi}} \exp\left\{-\frac{(t-\mu)^2}{2\sigma^2}\right\} \frac{dt}{\sigma},$$

so the pdf of $Y = \sigma X + \mu$ is

$$f_Y(y) = \frac{1}{\sqrt{2\pi\sigma^2}} \exp\left\{-\frac{(y-\mu)^2}{2\sigma^2}\right\}. \tag{2.15}$$

By (2.12)–(2.13), $\mathbb{E}[Y] = \mu$ and $\mathbb{V}[Y] = \sigma^2$. This is the **normal distribution**, written $Y \sim N(\mu, \sigma^2)$. ■

2.2 Multivariate Distributions

For events A, B with $P(B) > 0$,

$$P(A \mid B) := \frac{P(A \cap B)}{P(B)}$$

is the **conditional probability** of A given B. If

$$P(A \cap B) = P(A)P(B),$$

A and B are **independent**.

Similarly, for any $x, y \in \mathbb{R}$, we say random variables X, Y are independent if

$$P(X \le x, Y \le y) = P(X \le x)P(Y \le y). \tag{2.16}$$

In the discrete case (Problem 5), for all $a, b \in \mathbb{N}$,

$$P(X = a, Y = b) = P(X = a)\, P(Y = b). \tag{2.17}$$

In the continuous case, this is equivalent (w.p.1) to

$$f_{XY}(x, y) = f_X(x)\, f_Y(y), \tag{2.18}$$

where f_{XY} is the joint pdf satisfying

$$P(X \le x, Y \le y) = \int_{-\infty}^{x} \int_{-\infty}^{y} f_{XY}(s, t)\, ds\, dt.$$

For a function $h(X, Y)$,

$$\mathbb{E}_{XY}[h(X, Y)] := \begin{cases} \displaystyle\sum_{k=0}^{\infty} \sum_{l=0}^{\infty} h(k, l) P(X = k, Y = l), & \text{discrete} \\ \displaystyle\int_{-\infty}^{\infty} \int_{-\infty}^{\infty} h(x, y) f_{XY}(x, y)\, dx\, dy, & \text{continuous} \end{cases}$$

denotes the expectation.[4] With means μ_X, μ_Y,

$$\operatorname{cov}(X, Y) := \mathbb{E}_{XY}[(X - \mu_X)(Y - \mu_Y)]$$

is the **covariance**, and when variances are nonzero,

$$\rho(X, Y) := \frac{\operatorname{cov}(X, Y)}{\sqrt{\mathbb{V}[X]\mathbb{V}[Y]}}$$

is the **correlation coefficient**, always in $[-1, 1]$ (Problem 6). The **covariance matrix** is

$$\Sigma = \begin{bmatrix} \mathbb{V}[X] & \operatorname{cov}(X, Y) \\ \operatorname{cov}(X, Y) & \mathbb{V}[Y] \end{bmatrix} = \begin{bmatrix} \mathbb{V}[X] & \rho(X, Y)\sqrt{\mathbb{V}[X]\mathbb{V}[Y]} \\ \rho(X, Y)\sqrt{\mathbb{V}[X]\mathbb{V}[Y]} & \mathbb{V}[Y] \end{bmatrix}.$$

For a linear transform with $A \in \mathbb{R}^{2\times 2}$, $B \in \mathbb{R}^2$,

$$A\begin{bmatrix} X \\ Y \end{bmatrix} + B,$$

[4] We may write $\mathbb{E}_X[\cdot]$ or $\mathbb{E}_Y[\cdot]$ to emphasize which variable the expectation is taken over.

the mean and covariance are

$$A \begin{bmatrix} \mathbb{E}[X] \\ \mathbb{E}[Y] \end{bmatrix} + B, \qquad A \Sigma A^{\top} \tag{2.19}$$

(Problem 4).

Example 2.6 (Bivariate Normal) Let U, V be independent standard normals, so

$$f_{UV}(u, v) = f_U(u) f_V(v) = \frac{1}{2\pi} \exp\left(-\frac{u^2 + v^2}{2}\right).$$

Set

$$\sigma_X := \sqrt{\mathbb{V}[X]}, \quad \sigma_Y := \sqrt{\mathbb{V}[Y]}, \quad \mu_X := \mathbb{E}[X], \quad \mu_Y := \mathbb{E}[Y], \quad \rho_{XY} = \rho(X, Y),$$

and define

$$\begin{bmatrix} x \\ y \end{bmatrix} = A \begin{bmatrix} u \\ v \end{bmatrix} + \begin{bmatrix} \mu_X \\ \mu_Y \end{bmatrix}, \quad A = \begin{bmatrix} \sigma_X & 0 \\ \sigma_Y \rho_{XY} & \sigma_Y \sqrt{1 - \rho_{XY}^2} \end{bmatrix}.$$

Then

$$[x - \mu_X, y - \mu_Y](AA^{\top})^{-1} \begin{bmatrix} x - \mu_X \\ y - \mu_Y \end{bmatrix} = u^2 + v^2,$$

and

$$AA^{\top} = \begin{bmatrix} \sigma_X^2 & \sigma_X \sigma_Y \rho_{XY} \\ \sigma_X \sigma_Y \rho_{XY} & \sigma_Y^2 \end{bmatrix} =: \Sigma_{XY}.$$

As in Example 2.3, the Jacobian is

$$J = \det(A^{-1}) = \frac{1}{\sigma_X \sigma_Y \sqrt{1 - \rho_{XY}^2}} = \frac{1}{\sqrt{\det \Sigma_{XY}}}.$$

Thus

$$f_{XY}(x, y) = \frac{1}{2\pi \sqrt{\det \Sigma_{XY}}} \exp\left\{-\frac{1}{2}[x - \mu_X, y - \mu_Y] \Sigma_{XY}^{-1} \begin{bmatrix} x - \mu_X \\ y - \mu_Y \end{bmatrix}\right\},$$

with mean (μ_X, μ_Y) and covariance Σ_{XY}. ■

In general, if independent $U_1, \ldots, U_p$ are standard normal and

$$\begin{bmatrix} X_1 \\ \vdots \\ X_p \end{bmatrix} = A \begin{bmatrix} U_1 \\ \vdots \\ U_p \end{bmatrix} + \mu$$

for some $A \in \mathbb{R}^{p\times p}$ and $\mu \in \mathbb{R}^p$, then $[X_1, \ldots, X_p]^\top \sim N(\mu, \Sigma)$ with $\Sigma = AA^\top$.

A symmetric matrix $A \in \mathbb{R}^{p\times p}$ is **nonnegative definite** if $x^\top Ax \geq 0$ for all x and **positive definite** if $x^\top Ax > 0$ for all $x \neq 0$. A symmetric matrix is positive definite iff all eigenvalues are positive.

For any positive definite symmetric A, there exists a unique lower triangular L with positive diagonal entries such that $A = LL^\top$; this is the **Cholesky decomposition**.

Example 2.7 Covariance matrices are nonnegative definite. If $X_1, \ldots, X_p$ have mean $\mu_1, \ldots, \mu_p$ and covariance Σ, then for any $z \in \mathbb{R}^p$,

$$z^\top \Sigma z = \mathbb{E}\left[\left\{\sum_{i=1}^{p} z_i(X_i - \mu_i)\right\}^2\right] \geq 0.$$

Thus there is a unique lower triangular A with positive diagonal such that $\Sigma = AA^\top$. ■

For continuous X, Y, the joint density f_{XY} contains all information. Marginals are

$$f_X(x) = \int_{-\infty}^{\infty} f_{XY}(x, y)\, dy, \qquad f_Y(y) = \int_{-\infty}^{\infty} f_{XY}(x, y)\, dx,$$

and the conditional density is

$$f_{Y|X}(x \mid y) := \frac{f_{XY}(x, y)}{f_X(x)}, \qquad \int_{-\infty}^{\infty} f_{Y|X}(x \mid y)\, dy = 1,$$

so

$$f_{XY}(x, y) = f_X(x)\, f_{Y|X}(x \mid y).$$

Example 2.8 (Marginalization of Bivariate Normals) If $X \sim N(\mu_X, \sigma_X^2)$, $Y \sim N(\mu_Y, \sigma_Y^2)$ with correlation ρ, then

$$Y \mid (X = x) \sim N\left(\mu_Y + \rho\frac{\sigma_Y}{\sigma_X}(x - \mu_X),\ (1 - \rho^2)\sigma_Y^2\right) \tag{2.20}$$

(Problem 7). ■

If X, Y are independent, then $\rho(X, Y) = 0$ and $\text{cov}(X, Y) = 0$. For continuous variables,

$$\iint (x - \mu_X)(y - \mu_Y) f_X(x) f_Y(y)\, dx\, dy$$
$$= \Big(\int (x - \mu_X) f_X(x)\, dx\Big)\Big(\int (y - \mu_Y) f_Y(y)\, dy\Big) = 0.$$

The discrete case is analogous. However, zero covariance does not imply independence.

Example 2.9 Let X be standard normal and $Y = X^2$. Then X, Y are not independent, but $\text{cov}(X, Y) = 0$ because $\mathbb{E}[X] = \mathbb{E}[X^3] = 0$ and $\mathbb{E}[Y] = \mathbb{E}[X^2] = 1$:

$$\text{cov}(X, Y) = \text{cov}(X, X^2) = \mathbb{E}[X(X^2 - 1)] = \mathbb{E}[X^3] - \mathbb{E}[X] = 0.$$

■

For the normal family, the converse holds.

Proposition 2.1 *For a bivariate normal distribution, zero covariance is equivalent to independence.*

Proof If $\text{cov}(X, Y) = 0$, then

$$\Sigma = \text{diag}(\sigma_X^2, \sigma_Y^2), \quad \Sigma^{-1} = \text{diag}(\sigma_X^{-2}, \sigma_Y^{-2}),$$

and

$$f_{XY}(x, y) = \frac{1}{\sqrt{2\pi\sigma_X^2}} \exp\left(-\frac{(x - \mu_X)^2}{2\sigma_X^2}\right) \cdot \frac{1}{\sqrt{2\pi\sigma_Y^2}} \exp\left(-\frac{(y - \mu_Y)^2}{2\sigma_Y^2}\right),$$

which factors into $f_X(x) f_Y(y)$; hence X and Y are independent. ■

Finally, define the **characteristic function** of X by

$$\Psi_X(t) := \mathbb{E}[e^{iXt}].$$

If a pdf f_X exists,

$$\Psi_X(t) = \int_{-\infty}^{\infty} e^{ixt} f_X(x)\, dx,$$

a complex-valued function. Then

$$\Psi_X(0) = 1 \tag{2.21}$$

$$|\Psi_X(t)| \leq \mathbb{E}[|e^{iXt}|] = \mathbb{E}[1] = 1 \tag{2.22}$$

$$\Psi'_X(0) = i\,\mathbb{E}[X] \tag{2.23}$$

$$\Psi''_X(0) = -\,\mathbb{E}[X^2] \tag{2.24}$$

hold. For $X_1, \ldots, X_n$,

$$\Psi_{X_1,\ldots,X_n}(t_1, \ldots, t_n) := \mathbb{E}[e^{i(X_1t_1+\cdots+X_nt_n)}].$$

Example 2.10 (Normal Distribution) For $X \sim N(\mu, \sigma^2)$,

$$\Psi_X(t) = \int_{-\infty}^{\infty} e^{itx} \frac{1}{\sqrt{2\pi\sigma^2}} \exp\left(-\frac{(x-\mu)^2}{2\sigma^2}\right) dx$$
$$= \exp\left(i\mu t - \frac{\sigma^2 t^2}{2}\right) \tag{2.25}$$

(using completing-the-square and (2.5)). This satisfies (2.21)–(2.24) (Problem 8). ■

Characteristic functions are in one-to-one correspondence with distributions. Moreover, for independent $X_1, \ldots, X_n$,

$$\Psi_{X_1,\ldots,X_n}(t_1, \ldots, t_n) = \Psi_{X_1}(t_1) \cdots \Psi_{X_n}(t_n). \tag{2.26}$$

Example 2.11 (Binomial Distribution) If $X \in \{0, 1\}$ with $P(X = 1) = p$, then $\Psi_X(t) = 1 - p + pe^{it}$. Using (2.26), the characteristic function of the binomial variable Y in Example 2.1 is $\Psi_Y(t) = (1 - p + pe^{it})^n$, satisfying (2.21)–(2.24) (Problem 8). This also implies mean np and variance $np(1-p)$. ■

For random variables $X_1, X_2, \ldots$ and cdf $F_X(x) := P(X \leq x)$, we say X_n **converges in law** to X if $F_{X_n}(x_0) \to F_X(x_0)$ at all continuity points x_0 of F_X.

Example 2.12 (Central Limit Theorem) For X with mean zero and variance σ^2, $\Psi_X(t) = \exp(-\sigma^2 t^2/2 + o(t^2))$ near $t = 0$[5]. Let $X_1, \ldots, X_n$ be independent with mean μ and variance σ^2. Then

$$U_n := \frac{X_1 + \cdots + X_n - n\mu}{\sqrt{n}\sigma}$$

[5] $o(t^2)$ means $f(t)/t^2 \to 0$ as $t \to 0$.

has characteristic function

$$\Psi_{U_n}(t) = \left(1 - \frac{t^2}{2n} + o\left(\frac{1}{n}\right)\right)^n \to e^{-t^2/2},$$

the standard normal characteristic function; hence U_n converges in law to $N(0, 1)$.

2.3 Inverse Wishart Distribution

A distribution is said to have the **reproductive property** if sums of independent random variables with that distribution have the same distributional form (possibly with different parameters).

Proposition 2.2 (Reproductivity of χ^2) *If*

$$X \sim \chi_l^2, \quad Y \sim \chi_m^2, \quad X \perp\!\!\!\perp Y, \quad \textit{then} \quad X + Y \sim \chi_{l+m}^2.$$

Proof Using $\Psi_X(t) = (1 - 2it)^{-n/2}$ for $X \sim \chi_n^2$ (Problem 9) and independence,

$$\Psi_{X+Y}(t) = (1 - 2it)^{-l/2}(1 - 2it)^{-m/2} = (1 - 2it)^{-(l+m)/2}.$$

By uniqueness of characteristic functions, the claim follows. ■

Similarly, normals are reproductive.

Proposition 2.3 (Reproductivity of Normals) *If*

$$X \sim N(\mu_X, \sigma_X^2), \quad Y \sim N(\mu_Y, \sigma_Y^2), \quad X \perp\!\!\!\perp Y,$$

then $X + Y \sim N(\mu_X + \mu_Y,\ \sigma_X^2 + \sigma_Y^2)$.

Proof By (2.25) and independence,

$$\Psi_{X+Y}(t) = \mathbb{E}[e^{iXt}]\mathbb{E}[e^{iYt}] = \exp\left(i\mu_X t - \frac{\sigma_X^2 t^2}{2}\right)\exp\left(i\mu_Y t - \frac{\sigma_Y^2 t^2}{2}\right)$$

$$= \exp\left(i(\mu_X + \mu_Y)t - \frac{(\sigma_X^2 + \sigma_Y^2)t^2}{2}\right).$$

Uniqueness yields the result. ■

Let $X_1, \ldots, X_n \sim N(0, \Sigma)$ be independent p-variate normals ($0 \in \mathbb{R}^p$, $\Sigma \in \mathbb{R}^{p \times p}$). Define, for $a > 0$, the p-dimensional Gamma density

$$f_{p,a}(B) := \frac{1}{\Gamma_p(a)}(\det B)^{a - \frac{p+1}{2}} \exp\{-\mathrm{tr}(B)\}, \quad B > 0, \tag{2.27}$$

where B ranges over positive definite symmetric matrices. Here,

$$\Gamma_p(a) = \int_{B>0} (\det B)^{a-\frac{p+1}{2}} \exp\{-\mathrm{tr}(B)\}\,(dB) \tag{2.28}$$

is the p-dimensional Gamma function (an extension of 2.7); $(dB) = \prod_{i=1}^{p}\prod_{j=1}^{i} db_{i,j}$ for the lower triangular entries.

For $p = 1$, (2.7) gives $\Gamma_1(a) = \Gamma(a)$. For $p = 2$, one shows

$$\Gamma_2(a) = \sqrt{\pi}\,\Gamma(a)\,\Gamma\left(a - \frac{1}{2}\right) \tag{2.29}$$

(Problem 11). In general, we have the following proposition:

Proposition 2.4

$$\Gamma_p(\alpha) = \pi^{p(p-1)/4} \prod_{j=1}^{p} \Gamma\left(\alpha + \frac{1-j}{2}\right), \qquad \alpha > \frac{p-1}{2}.$$

The proof is given in the Appendix.

Next, put $a = \nu/2$, let A be the lower triangular Cholesky factor of Σ ($\Sigma = AA^\top$) with positive diagonal, set $S := (\sqrt{2}A)^{-1}$, and change variables $B = SMS^\top$. This yields the density on positive definite M:

$$f_{\Sigma,\nu}(M) = \frac{(\det \Sigma)^{-\nu/2}}{2^{\nu p/2}\Gamma_p(\nu/2)} (\det M)^{(\nu-p-1)/2} \exp\left\{-\frac{1}{2}\mathrm{tr}(\Sigma^{-1}M)\right\}. \tag{2.30}$$

When $\nu = n \geq p$ is an integer, (2.30) is the density of $M = X_1X_1^\top + \cdots + X_nX_n^\top$, called the **Wishart distribution** [1]. We write $M \sim W(\Sigma, \nu)$, allowing real $\nu > p - 1$.

The distribution of $U := M^{-1}$ is the **inverse Wishart**; with $\Lambda := \Sigma^{-1}$, write $U \sim W^{-1}(\Lambda, \nu)$, having density

$$f_{\nu,\Lambda}(U) = \frac{(\det \Lambda)^{\nu/2}}{2^{\nu p/2}\Gamma_p(\nu/2)} (\det U)^{-(\nu+p+1)/2} \exp\left\{-\frac{1}{2}\mathrm{tr}(\Lambda U^{-1})\right\}. \tag{2.31}$$

2.4 Maximum Likelihood Estimation

When a normal distribution is assumed, the mean $\mu \in \mathbb{R}$ and variance $\sigma^2 > 0$ may be known or unknown. In the latter case, from observations $X_1, \ldots, X_n \sim N(\mu, \sigma^2)$ we often estimate

$$\bar{X} := \frac{1}{n}\sum_{i=1}^{n} X_i$$

for μ and

$$S^2 := \frac{1}{n}\sum_{i=1}^{n}(X_i - \bar{X})^2$$

for σ^2. Many estimation methods exist. Here we explain **maximum likelihood**.

Given $X_1 = x_1, \ldots, X_n = x_n$, the likelihood is $p(x_1, \ldots, x_n \mid \theta)$. If the X_i are independent, this factors as $\prod_{i=1}^{n} p(x_i \mid \theta)$. Viewing it as a function of θ, a maximizer $\hat{\theta} = \hat{\theta}(x_1, \ldots, x_n)$ is the **maximum likelihood estimate** (MLE).

Example 2.13 Two baseball teams play three games (no ties). There are $2^3 = 8$ possible sequences. Let team A win each game with probability θ (team B with $1 - \theta$). For game i, set $X_i = 1$ if A wins, else 0. Assuming independence, for (x_1, x_2, x_3),

$$p(x_1x_2x_3 \mid \theta) = \theta^k(1-\theta)^{3-k},$$

with $k = x_1 + x_2 + x_3$. For $(1, 1, 0)$, $p(110 \mid \theta) = \theta^2(1-\theta)$ (Table 2.1, column 2). Maximizing $p(1, 1, 0 \mid \theta)$ is equivalent to maximizing $L = \log\{\theta^2(1-\theta)\}$:

$$\frac{dL}{d\theta} = \frac{2}{\theta} - \frac{1}{1-\theta} = 0 \iff \theta = \frac{2}{3}.$$

Hence $\hat{\theta} = 2/3$ and $p(1, 1, 0 \mid \hat{\theta}) = \frac{4}{27}$. ∎

Below we assume n i.i.d. observations from the same distribution.

Example 2.14 Let X take values in $\{1, \ldots, \alpha\}$. If in $X_1 = x_1, \ldots, X_n = x_n$ the counts of $1, \ldots, \alpha$ are $k_1, \ldots, k_\alpha$, the likelihood

Table 2.1 Likelihood and maximum likelihood estimator

(x_1, x_2, x_3)	$p(x_1x_2x_3 \mid \theta)$	$p(x_1x_2x_3 \mid \theta = 2/3)$	$\hat{\theta}(x_1, x_2, x_3)$	$p(x_1x_2x_3 \mid \hat{\theta}(x_1, x_2, x_3))$
(0, 0, 0)	$(1-\theta)^3$	1/27	0	1
(1, 0, 0)	$(1-\theta)^2\theta$	2/27	1/3	8/27
(0, 1, 0)	$(1-\theta)^2\theta$	2/27	1/3	8/27
(0, 0, 1)	$(1-\theta)^2\theta$	2/27	1/3	8/27
(0, 1, 1)	$(1-\theta)\theta^2$	4/27	2/3	8/27
(1, 0, 1)	$(1-\theta)\theta^2$	4/27	2/3	8/27
(1, 1, 0)	$(1-\theta)\theta^2$	4/27	2/3	8/27
(1, 1, 1)	θ^3	8/27	1	1

$$\prod_{i=1}^{n} p(x_i \mid \theta) = \theta_1^{k_1} \cdots \theta_\alpha^{k_\alpha}$$

is maximized at the MLEs

$$\hat{\theta}_1 = \frac{k_1}{n}, \ \ldots, \ \hat{\theta}_\alpha = \frac{k_\alpha}{n}$$

(Problem 12). ■

For $\epsilon > 0$, we say X_n **converges in probability** to a if $P(|X_n - a| < \epsilon) \to 1$. In Example 2.14, $\hat{\theta}_j = k_j/n \to \theta_j$ in probability. This is the **Law of Large Numbers**.

Example 2.15 Given n data pairs $(x_i, y_i) \in \mathbb{R}^2$, fit a line $y = \beta_0 + \beta_1 x$. Assume $Y \mid (X = x) \sim N(\beta_0 + \beta_1 x, \sigma^2)$. The likelihood is

$$\prod_{i=1}^{n} p(y_i \mid x_i, \beta_0, \beta_1, \sigma^2) = (2\pi\sigma^2)^{-n/2} \exp\left\{-\frac{1}{2\sigma^2}\sum_{i=1}^{n}(y_i - \beta_0 - x_i\beta_1)^2\right\},$$

which is

$$L = -\frac{n}{2}\log(2\pi\sigma^2) + \frac{1}{2\sigma^2}\sum_{i=1}^{n}(y_i - \beta_0 - x_i\beta_1)^2.$$

Differentiating w.r.t. $\beta_0, \beta_1, \sigma^2$ yields three equations. Solving the first two (equivalently least squares) gives $\hat{\beta}_0, \hat{\beta}_1$ (Problem 13); substituting into the third gives

$$\hat{\sigma}^2 = \frac{1}{n}\sum_{i=1}^{n}(y_i - \hat{\beta}_0 - x_i\hat{\beta}_1)^2. \tag{2.32}$$

Plugging back into L yields

$$L = -\frac{n}{2}\log(2\pi\hat{\sigma}^2 e) \tag{2.33}$$

(Problem 14). ■

2.5 Statistical Testing

A teacher claims: "Boys grow 5 cm from the first to the second year of middle school. Suppose growth is normal with known variance $\sigma^2 = 9$ and unknown mean μ. From n observed growth values $x_1, \ldots, x_n$, we will **statistically test** this claim.

Note we are not asking whether $\overline{x} = \frac{1}{n}\sum x_i$ equals μ. Since there are many boys nationwide, we treat $x_1, \ldots, x_n$ as one random sample from $N(\mu, \sigma^2)$, tolerating some error in estimation/testing.

We take the hypothesis $\mu = 5$ as the **null hypothesis**. If assuming it leads to a contradiction with data, we reject it; otherwise we do not reject it. The competing claim $\mu \neq 5$ is the **alternative hypothesis**. Testing assesses whether the data provide sufficient evidence for the alternative.

There are two kinds of testing errors:

Type I error: Reject the null though it is true.
Type II error: Fail to reject the null though it is false.

Examples include medical diagnosis (healthy vs. ill), quality control (good vs. defective), and policing (innocent vs. guilty).

We set a significance level α (e.g., 5%) for Type I error and aim to minimize β (the Type II error rate) or equivalently maximize the **power** $1 - \beta$. Multiple testing procedures may be available; for fixed α, a test with larger power $1 - \beta$ is preferable. We can visualize performance via the **ROC curve** (Receiver Operating Characteristic); a larger **AUC** (Area Under Curve) indicates a better test.

Tests on the same ROC curve use the same method but different thresholds (e.g., fever threshold 37°C vs. 38°C). Different methods (e.g., a saliva-based assay) yield different ROC curves; comparing AUCs helps select the better method (Fig. 2.2).

Fixing α means we tolerate Type I errors up to probability α. If $X_i \sim N(\mu, \sigma^2)$, then

$$\frac{1}{n}\sum_{i=1}^{n} X_i \sim N\left(\mu, \frac{\sigma^2}{n}\right).$$

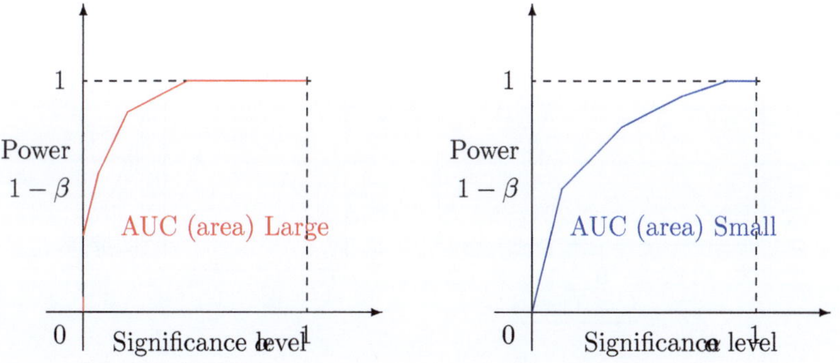

Fig. 2.2 ROC curves: a test with a large AUC versus a test with a small AUC

Indeed,

$$\mathbb{E}\left[\frac{1}{n}\sum_{i=1}^{n} X_i\right] = \mu,$$

$$\mathbb{V}(\overline{X}) = \frac{1}{n^2}\sum_{i=1}^{n}\sum_{j=1}^{n}\operatorname{cov}(X_i, X_j) = \frac{1}{n^2}\cdot n\sigma^2 = \frac{\sigma^2}{n},$$

using independence for $i \neq j$. Hence

$$\frac{\sqrt{n}(\overline{X}-\mu)}{\sigma} \sim N(0,1) \tag{2.34}$$

(Problem 15).

A standard two-sided test of $H_0 : \mu = 5$ uses

$$Z := \frac{\sqrt{n}(\overline{X}-5)}{3},$$

which is standard normal under H_0. We reject H_0 if Z falls in the two tails with probability $\alpha/2$ each. The distribution of the test statistic under H_0 is the **null distribution** (Fig. 2.3).

In our example, a two-sided test takes $H_0 : \mu = 5$ vs. $H_1 : \mu \neq 5$, placing rejection regions of size $\alpha/2$ in both tails (Fig. 2.4a). One-sided tests use a single tail: $H_1 : \mu > 5$ (right-sided, Fig. 2.4c) or $H_1 : \mu < 5$ (left-sided, Fig. 2.4b). In Chap. 4, tests of (conditional) independence are one-sided.

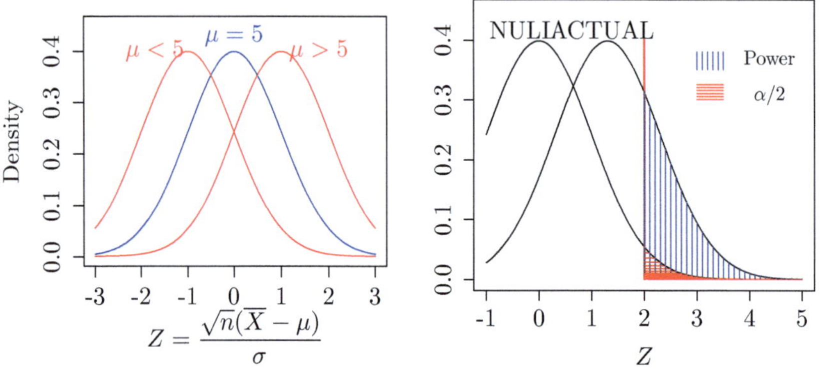

Fig. 2.3 The distribution of Z varies with the true mean (left). Under the null, Z in the rejection region (right tail) constitutes a Type I error; under the true (shifted) distribution, the blue vertical area indicates the power

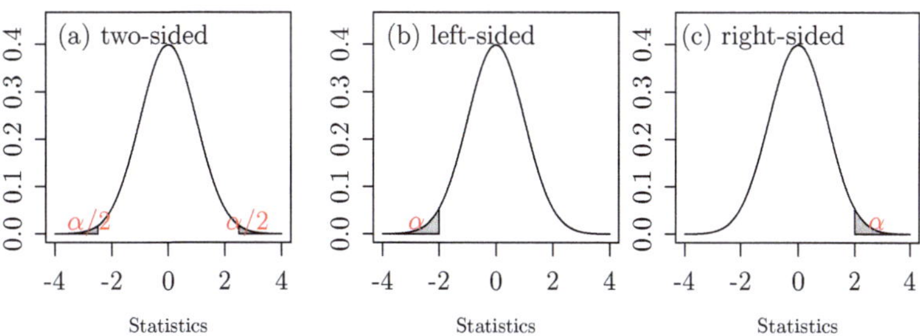

Fig. 2.4 Two-sided and one-sided tests

Example 2.16 For $n = 10$ growth values

```
5.0, 4.9, 5.3, 5.2, 3.9, 4.5, 6.1, 4.8, 5.2, 4.3
```

and $\alpha = 0.05$ two-sided, the following R code shows H_0 is not rejected:

```
x <- c(5.0, 4.9, 5.3, 5.2, 3.9, 4.5, 6.1, 4.8, 5.2, 4.3)
mu <- 5
sigma <- 3
n <- length(x)
sqrt(n) * (mean(x)-mu) / sigma    ## value of Z
qnorm(0.025)                      ## lower critical value
qnorm(0.975)                      ## upper critical value
```

```
> sqrt(n)*(mean(x)-mu)/sigma
[1] -0.0843274
> qnorm(0.025)
[1] -1.959964
> qnorm(0.975)
[1] 1.959964
```

■

Appendix

We identify a positive definite symmetric matrix or a lower triangular matrix with the vector consisting of its $p(p+1)/2$ lower triangular elements. The Jacobians of the following transformations are known. For Wishart and inverse Wishart results and the proof of Proposition 5, see Chapter 2 of [17].

Proposition 2.5 *For a positive definite symmetric $X \in \mathbb{R}^{p \times p}$:*

1. *For nonsingular $S \in \mathbb{R}^{p \times p}$, with $X = SYS^\top$ and $T = S^\top S$,*

$$(dX) = (\det T)^{\frac{p+1}{2}}(dY), \quad \det X = (\det T)(\det Y), \quad \operatorname{tr}(X) = \operatorname{tr}(TY).$$

2. *For* $X = Y^{-1}$, $(dX) = (\det Y)^{-(p+1)}(dY)$.
3. *For* $X = YY^\top$ *with lower triangular* $Y = (y_{ij})$ *having positive diagonal,*

$$(dX) = 2^p \prod_{i=1}^{p} y_{ii}^{p+1-i} \cdot \prod_{i=1}^{p}\prod_{j=1}^{i} dy_{ij}, \quad \det X = \prod_{i=1}^{p} y_{ii}^2, \quad \mathrm{tr}(X) = \sum_{i=1}^{p}\sum_{j=1}^{i} y_{ij}^2.$$

Using Proposition 2.5, we prove (2.30), (2.31), and Proposition 2.4, in that order.

Proofs of (2.30) and (2.31)

From Proposition 2.5(1), with $T = (2\Sigma)^{-1}$ and $Y = M$,

$$\begin{aligned}(dB) &= (\det(2\Sigma)^{-1})^{\frac{p+1}{2}}(dM) = 2^{-\frac{p(p+1)}{2}}(\det\Sigma)^{-\frac{p+1}{2}}(dM),\\ \det(B) &= 2^{-p}(\det\Sigma)^{-1}\det(M),\\ \mathrm{tr}(B) &= \tfrac{1}{2}\,\mathrm{tr}(\Sigma^{-1}M).\end{aligned}$$

Substituting into (2.27) gives (2.30).

For (2.31), by Proposition 2.5(2), $(dM) = (\det U)^{-(p+1)}(dU)$. Substitute $M = U^{-1}$ into (2.30) and multiply by $(\det U)^{-(p+1)}$.

Proof of Proposition 2.4

Via Cholesky, any $B > 0$ can be written $B = UU^\top$ with lower triangular U having positive diagonal. By Proposition 2.5(3),

$$\begin{aligned}\Gamma_p(a) &= 2^p \int\cdots\int \exp\left(-\sum_{i=1}^{p}\sum_{j=1}^{i} u_{ij}^2\right)\left(\prod_{i=1}^{p} u_{ii}^2\right)^{a-(p+1)/2} \prod_{i=1}^{p} u_{ii}^{p+1-i} \prod_{i=1}^{p}\prod_{j=1}^{i} du_{ij}\\ &= \left\{\int_{-\infty}^{\infty} e^{-s^2} ds\right\}^{p(p-1)/2} \prod_{i=1}^{p}\left\{\int_0^{\infty} e^{-u_{ii}^2}(u_{ii}^2)^{a-i/2}\, 2\, du_{ii}\right\}\\ &= \pi^{p(p-1)/4}\prod_{i=1}^{p}\Gamma\left(a - \frac{i-1}{2}\right),\end{aligned}$$

which is the claimed formula (Problem 17). Here B-integration corresponds to integrating over $u_{ii} \geq 0$ and $u_{ij} \in \mathbb{R}$ for $i \neq j$, with $(dB) = \prod_{i=1}^{p}\prod_{j=1}^{i} du_{ij}$.

The last step uses the normal integral with $\sigma^2 = 1/2$ and the Gamma density (2.9) with a replaced by $a - (p-1)/2$. ■

Problems 1–18

1. Derive (2.1) and (2.2) from the three axioms of probability. *Hint:* Regarding (2.2), note that

$$\begin{aligned} A \cup B &= (A - A \cap B) \cup (B - A \cap B) \cup A \cap B, \\ A &= (A - A \cap B) \cup A \cap B, \\ B &= (B - A \cap B) \cup A \cap B \end{aligned}$$

   where in each case the event on the left-hand side is written as the union of disjoint events on the right-hand side, so each term can be replaced by its probability. From this, the desired relation follows.
2. Show the last two equalities of (2.6).
3. Show (2.8) and $\Gamma(1/2) = \sqrt{\pi}$. *Hint:* In (2.7) with $\alpha = 1/2$, perform the substitution $t = x^2/2$ in the integral.
4. Prove equations (2.12), (2.13), (2.14), and (2.19). You may assume continuous random variables.
5. For discrete random variables, show that the validity of (2.16) for all $x, y \in \mathbb{N}$ is equivalent to the validity of (2.17) for all $a, b \in \mathbb{N}$. *Hint:* Since X and Y take at most countably many values, we may assume they take values in the nonnegative integers. Using

$$P(X \le x, Y \le y) = \sum_{a=0}^{x} \sum_{b=0}^{y} P(X = a, Y = b),$$

   one can prove (2.17) $\Rightarrow$ (2.16). For the converse, substitute $(x, y) = (a, b), (a, b-1)$ into (2.16) to obtain a relation. Replace a by $a - 1$ to obtain another relation, and combine them to obtain the final relation.
6. Assuming $\mathbb{V}[Y] > 0$, let $Z := X - \dfrac{cov(X, Y)}{\mathbb{V}[Y]} Y$. Show that $cov(Y, Z) = 0$ and that

$$\mathbb{V}[X] \ge \frac{cov(X, Y)^2}{\mathbb{V}[Y]}.$$

   Hint: Substitute $X = Z + \frac{cov(X,Y)}{\mathbb{V}[Y]} Y$ into $\mathbb{V}[X]$.

7. Show (2.20). *Hint:* Transform as follows:

$$\frac{1}{2\pi\sigma_X\sigma_Y\sqrt{1-\rho^2}}$$
$$\cdot \exp\left\{-\frac{1}{2(1-\rho^2)}\left[\left(\frac{x-\mu_X}{\sigma_X}\right)^2 - 2\rho\left(\frac{x-\mu_X}{\sigma_X}\right)\left(\frac{y-\mu_Y}{\sigma_Y}\right)+\left(\frac{y-\mu_Y}{\sigma_Y}\right)^2\right]\right\}$$
$$= \frac{1}{\sqrt{2\pi\sigma_X^2}}\exp\left\{-\frac{1}{2}\left(\frac{x-\mu_X}{\sigma_X}\right)^2\right\}$$
$$\cdot \frac{1}{\sqrt{2\pi\sigma_Y^2(1-\rho^2)}}\exp\left\{-\frac{1}{2(1-\rho^2)}\left(\frac{y-\mu_Y}{\sigma_Y}-\rho\frac{x-\mu_X}{\sigma_X}\right)^2\right\}.$$

8. In Examples 2.10 and 2.11, show that the obtained characteristic functions satisfy (2.21)–(2.24). *Hint:* For Example 2.11, use

$$\Psi_X'(t) = inp(1-p+pe^{it})^{n-1}e^{it},$$
$$\Psi_X''(t) = -npe^{it}\left(1-p+pe^{it}\right)^{n-1} - np^2(n-1)e^{2it}\left(1-p+pe^{it}\right)^{n-2}.$$

9. For $X \sim \chi_n^2$, show that $\Psi_X(t) = (1-2it)^{-n/2}$. *Hint:* First derive

$$\Psi_X(t) = \frac{1}{2^{n/2}\Gamma(n/2)}\int_0^\infty x^{n/2-1}\exp\left(-\frac{1-2it}{2}x\right)dx,$$

and then let $y := (1-2it)x$.

10. Based on the central limit theorem, suppose you wish to generate a standard normal random variable by tossing a coin n times (with n sufficiently large). How can you construct such a random variable from the coin toss results? Also, generate such random variables 1000 times and output a histogram of their frequencies in R. *Hint:* Let $X_i = 1$ if heads and $X_i = 0$ if tails. Then $S_n = \sum_{i=1}^n X_i$ has mean $n/2$ and variance $n/4$.
11. Consider the derivation of (2.29).

    (a) Derive

$$J = \det\begin{bmatrix} \frac{\partial b_{11}}{\partial u} & \frac{\partial b_{11}}{\partial v} & \frac{\partial b_{11}}{\partial \theta} \\ \frac{\partial b_{22}}{\partial u} & \frac{\partial b_{22}}{\partial v} & \frac{\partial b_{22}}{\partial \theta} \\ \frac{\partial b_{12}}{\partial u} & \frac{\partial b_{12}}{\partial v} & \frac{\partial b_{12}}{\partial \theta}\end{bmatrix} = \det\begin{bmatrix} 1 & 0 & 0 \\ 0 & 1 & 0 \\ \frac{\partial b_{12}}{\partial u} & \frac{\partial b_{12}}{\partial v} & \sqrt{uv}\sin\theta\end{bmatrix} = \sqrt{uv}\sin\theta.$$

(b) Show

$$\int_0^{\pi} \sin^{2(a-1)} \theta d\theta = \frac{\Gamma(a-\frac{1}{2})\Gamma(\frac{1}{2})}{\Gamma(a)}.$$

Hint: Set $t = \cos^2\theta$ and use substitution. Since the integrand is even, halve the interval. Note that $\frac{dt}{d\theta} = -2\cos\theta\sin\theta = -2(1-t)^{1/2}t^{1/2}$. Then apply (1.11).

(c) Show (2.29).

12. Derive the maximum likelihood estimator in Example 2.14. *Hint:* The MLE maximizes ℓ under the constraint $\sum p_i = 1$. Consider

$$\sum_{i=1}^{\alpha} k_i \log p_i + \lambda\left(1 - \sum_{i=1}^{\alpha} p_i\right),$$

differentiate with respect to p_j, set equal to 0, express p_j in terms of λ, and use $\sum p_i = 1$ to solve for λ.

13. In Example 2.15, show that the solutions for $\hat{\beta}_0, \hat{\beta}_1$ are

$$\hat{\beta}_1 = \frac{\sum_{i=1}^{n}(x_i-\bar{x})(y_i-\bar{y})}{\sum_{i=1}^{n}(x_i-\bar{x})^2}, \qquad \hat{\beta}_0 = \bar{y} - \bar{x}\hat{\beta}_1.$$

First solve the simultaneous equations $\frac{\partial L}{\partial \beta_0} = 0$, $\frac{\partial L}{\partial \beta_1} = 0$. Begin by assuming $\bar{x} = \bar{y} = 0$, and then replace (x_i, y_i) by $(x_i - \bar{x}, y_i - \bar{y})$ to show that the slope does not change. Use this to obtain $\hat{\beta}_1$, and then determine $\hat{\beta}_0$ from $\bar{x}, \bar{y}, \hat{\beta}_1$.

14. Show (2.32) and (2.33).
15. Prove (2.34).
16. Regarding Example 2.16:

(a) Generate height data for ten junior high school students from $N(5, 3^2)$, set $\alpha = 0.05$, and perform a statistical test using R.

(b) Generate height data for ten junior high school students from $N(7, 3^2)$, and check whether the test statistic Z falls in the rejection region. Repeat this operation 1000 times for $\alpha = 0.05$, and estimate the detection rate. Also perform the same for $\alpha = 0.01$.

17. In the proof of Proposition 2.4 in the Appendix, explain why the five equalities in the derivation of $\Gamma_p(a)$ hold.

18. Let Y_p be a lower triangular matrix with all positive diagonal entries, and $X_p = Y_p Y_p^\top$. Show that

$$X_p = \begin{bmatrix} Y_{p-1} & 0 \\ y_{p-1} & y_{pp} \end{bmatrix} \begin{bmatrix} Y_{p-1}^\top & y_{p-1}^\top \\ 0 & y_{pp} \end{bmatrix} = \begin{bmatrix} X_{p-1} & Y_{p-1} y_{p-1}^\top \\ y_{p-1} Y_{p-1}^\top & y_{p-1} y_{p-1}^\top + y_{pp}^2 \end{bmatrix}.$$

Using this, prove part 3 of Proposition 2.5.

(a) Show that the Jacobian matrix of $\tilde{X}_p$ (consisting of the lower triangular entries of X_{p-1}, $y_{p-1} Y_{p-1}^\top$, and $y_{p-1} y_{p-1}^\top + y_{pp}^2$) with respect to $\tilde{Y}_p$ (the lower triangular entries of Y_p) can be written as

$$\frac{\partial \tilde{X}_p}{\partial \tilde{Y}_p} = \begin{bmatrix} \frac{\partial \tilde{X}_{p-1}}{\partial \tilde{Y}_{p-1}} & * & * \\ 0 & Y_{p-1}^\top & * \\ 0 & 0 & 2y_{p,p} \end{bmatrix}.$$

(b) Compute $\frac{\partial \tilde{X}_1}{\partial \tilde{Y}_1}$. Assuming that $\det\left(\frac{\partial \tilde{X}_{p-1}}{\partial \tilde{Y}_{p-1}}\right) = 2^{p-1} \prod_{i=1}^{p-1} y_{ii}^{p-i}$ holds, show that $\det\left(\frac{\partial \tilde{X}_p}{\partial \tilde{Y}_p}\right) = 2^p \prod_{i=1}^{p} y_{ii}^{p+1-i}$. *Hint:* Use $\det(Y_{p-1}) = \prod_{i=1}^{p-1} y_{ii}$.

(c) Find X_1. Then, assuming that $\operatorname{tr}(X_{p-1}) = \sum_{i=1}^{p-1} \sum_{j=1}^{i} y_{i,j}^2$ holds, show that $\operatorname{tr}(X_p) = \sum_{i=1}^{p} \sum_{j=1}^{i} y_{i,j}^2$. *Hint:* Use $y_{p-1} = [y_{p,1}, \ldots, y_{p,p-1}]$.

Chapter 3
Graphical Models

In this chapter, we study graphical models, which express conditional independence among random variables. A graphical model is a framework that visually represents probabilistic structure using either undirected graphs or directed acyclic graphs (DAGs).

We begin by introducing the graphoid axioms—symmetry, decomposition, weak union, and contraction—as the basic properties that conditional independence should satisfy and explain their meanings and roles. Next, we define separation in undirected graphs and d-separation in DAGs and formalize how these correspond to conditional independence.

We then define Markov networks based on undirected graphs and Bayesian networks based on DAGs, illustrating the differences and characteristics of the probabilistic structures they can represent. Finally, we introduce the Chow–Liu algorithm and demonstrate its usefulness as a method for constructing approximate distributions using mutual information. By using this algorithm to construct a maximum spanning tree of an undirected graph, we obtain an approximation that is optimal in terms of Kullback–Leibler divergence.

3.1 Conditional Independence

As an extension of independence, we have **conditional independence**.

When a random variable Z takes the value $Z = z \in \mathbb{R}$, we say random variables X and Y are independent given $Z = z$ if, for any $x, y \in \mathbb{R}$,

$$P(X \leq x, Y \leq y \mid Z = z) = P(X \leq x \mid Z = z)\, P(Y \leq y \mid Z = z).$$

J. Suzuki, *Graphical Models and Causal Discovery with Python*,
https://doi.org/10.1007/978-981-95-5308-2_3

If this holds for $z \in \mathbb{R}$ with probability one, we say that X and Y are conditionally independent given Z and write $X \perp\!\!\!\perp Y \mid Z$ (if X and Y are simply independent, we write $X \perp\!\!\!\perp Y$).

Accordingly, in the discrete and continuous cases we have, with probability one,

$$P(X = x, Y = y \mid Z = z) = P(X = x \mid Z = z)\, P(Y = y \mid Z = z),$$

$$f_{XY|Z}(x, y \mid z) = f_{X|Z}(x \mid z)\, f_{Y|Z}(y \mid z),$$

where $f_{XYZ}(x, y, z)$ denotes the joint density of X, Y, Z, and

$$f_{XY|Z}(x, y \mid z) := \frac{f_{XYZ}(x, y, z)}{f_Z(z)}.$$

Equivalently, one may write

$$P(X = x, Y = y, Z = z)\, P(Z = z) = P(X = x, Z = z)\, P(Y = y, Z = z),$$

$$f_{XYZ}(x, y, z)\, f_Z(z) = f_{XZ}(x, z)\, f_{YZ}(y, z).$$

Conditional independence can also be defined for multiple random variables, e.g.,

$$\{X, Y\} \perp\!\!\!\perp Z \mid \{U, V\}, \tag{3.1}$$

meaning "$\{X, Y\}$ and Z are conditionally independent given $\{U, V\}$." When representing multiple variables such as X, Y or U, V, we enclose them in braces $\{\cdot\}$.

Below, we sometimes suppress explicit mention of the argument, writing $P(X)$, f_X instead of $P(X = x)$, $f_X(x)$ when the value is clear from context.

Example 3.1 (Conditional Independence) Let the sample space be $\Omega = \{1, 2, 3, 4, 5, 6\}$, and consider events $A = \{1\}$, $B = \{1, 2\}$, $C = \{1, 2, 3\}$. Since

$$P(A \cap C \mid B) = P(A \mid B)\, P(C \mid B) = P(\{1\} \mid \{1, 2\}),$$

the events A and C are conditionally independent given B.

Example 3.2 (Conditional Independence and Its Graphical Representation) For discrete random variables X, Y, Z, if the joint distribution can be written as

$$P(X, Y, Z) = \frac{P(X, Z)\, P(Y, Z)}{P(Z)}, \tag{3.2}$$

then $X \perp\!\!\!\perp Y \mid Z$ holds. Indeed, for a specific value $Z = z$,

$$P(X, Y \mid Z = z) = P(X \mid Z = z)\, P(Y \mid Z = z),$$

so X and Y are independent given $Z = z$.

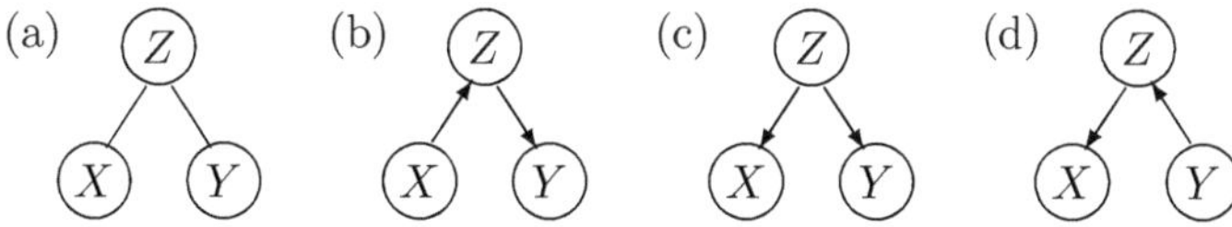

Fig. 3.1 The conditional independence $X \perp\!\!\!\perp Y \mid Z$ represented by the undirected graph (**a**) and the directed graphs (**b**), (**c**), and (**d**)

Equation (3.2) can be expressed in three ways:

$$P(X)\,P(Z \mid X)\,P(Y \mid Z) = P(Z)\,P(X \mid Z)\,P(Y \mid Z) = P(Y)\,P(Z \mid Y)\,P(X \mid Z). \tag{3.3}$$

These three factorizations are often represented by the undirected graph in Fig. 3.1a and the directed (acyclic) graphs in Fig. 3.1b,c,d. ∎

For conditional independence, the following proposition holds. These are the "graphoid axioms," important properties in the analysis of graphical models and cited later in the book.

Proposition 3.1 (Pearl–Paz [19]) *Let A, B, C, D be pairwise disjoint subsets of $\{1, \ldots, p\}$. Then*

$$X_A \perp\!\!\!\perp X_B \mid X_C \iff X_B \perp\!\!\!\perp X_A \mid X_C, \tag{3.4}$$

$$X_A \perp\!\!\!\perp X_{B\cup D} \mid X_C \Longrightarrow X_A \perp\!\!\!\perp X_B \mid X_C \text{ and } X_A \perp\!\!\!\perp X_D \mid X_C, \tag{3.5}$$

$$X_A \perp\!\!\!\perp X_{B\cup D} \mid X_C \Longrightarrow X_A \perp\!\!\!\perp X_B \mid X_{C\cup D}, \tag{3.6}$$

$$X_A \perp\!\!\!\perp X_B \mid X_C \text{ and } X_A \perp\!\!\!\perp X_D \mid X_{B\cup C} \Longrightarrow X_A \perp\!\!\!\perp X_{B\cup D} \mid X_C \tag{3.7}$$

hold.

In Proposition 3.1, if $A = \{1, 2\}$ and $B = \{2, 4\}$, then $X_A = \{X_1, X_2\}$ and $X_{A\cup B} = \{X_1, X_2, X_4\}$. The properties (3.4)–(3.7) are called **symmetry**, **decomposition**, **weak union**, and **contraction**, respectively.

Proof Symmetry follows from

$$P(X_A, X_B \mid X_C) = P(X_A \mid X_C)P(X_B \mid X_C) = P(X_B \mid X_C)P(X_A \mid X_C).$$

For decomposition, marginalizing $X_A \perp\!\!\!\perp X_{B\cup D} \mid X_C$ over X_D yields $X_A \perp\!\!\!\perp X_B \mid X_C$, and marginalizing over X_B yields $X_A \perp\!\!\!\perp X_D \mid X_C$. For weak union, from the assumption we have $X_A \perp\!\!\!\perp X_{B\cup D} \mid X_C$ and by decomposition also $X_A \perp\!\!\!\perp X_D \mid X_C$. Hence

$$P(X_A \mid X_C) = P(X_A \mid X_{B\cup C\cup D}), \qquad P(X_A \mid X_C) = P(X_A \mid X_{C\cup D}),$$

which implies

$$P(X_A \mid X_{B\cup C\cup D}) = P(X_A \mid X_{C\cup D}),$$

i.e., weak union holds. For contraction, the assumptions

$$X_A \perp\!\!\!\perp X_B \mid X_C, \qquad X_A \perp\!\!\!\perp X_D \mid X_{B\cup C}$$

are equivalent to

$$P(X_A \mid X_C) = P(X_A \mid X_{B\cup C}), \qquad P(X_A \mid X_{B\cup C}) = P(X_A \mid X_{B\cup C\cup D}),$$

and hence

$$P(X_A \mid X_C) = P(X_A \mid X_{B\cup C\cup D}),$$

which proves contraction. ■

3.2 Separation in Graphs

In this book, conditional independence among random variables is represented using undirected graphs and directed acyclic graphs. We introduce the notion of separation needed for this understanding.

Prepare p vertices $1, \ldots, p$, and construct a graph G by connecting them with edges. First, fix the **vertex set** $V = \{1, \ldots, p\}$. We consider undirected graphs consisting only of undirected edges and directed graphs consisting only of directed edges.

In an **undirected graph**, an undirected edge connecting vertices i, j is denoted by the set $\{i, j\}$. Since it is a set, $\{i, j\} = \{j, i\}$. The **edge set** E is defined as a subset of $\mathcal{E} := \{\{i, j\} \mid i, j \in V,\ i \neq j\}$. If $\{i, j\} \in E$, we say there exists a **path** from i to j. If $i, j \in V$ are connected by a path and $\{j, k\} \in E$, then there exists a path from i to k.

Example 3.3 (Undirected Graphs) For the vertex set $V = \{1, 2, 3\}$, there are eight undirected graphs $G = (V, E)$ as shown in Fig. 3.2. Their edge sets E are

$$\{\},\ \ \{\{1,2\}\},\ \ \{\{2,3\}\},\ \ \{\{3,1\}\},\ \ \{\{3,1\},\{3,2\}\},$$
$$\{\{1,2\},\{1,3\}\},\ \ \{\{2,3\},\{2,1\}\},\ \ \{\{2,3\},\{3,1\},\{1,2\}\}.$$

In (e), (f), and (g), there is no path from 1 to 2, from 2 to 3, and from 1 to 3, respectively. In (h), there exists a cycle (a path that starts at a vertex and returns to it). ■

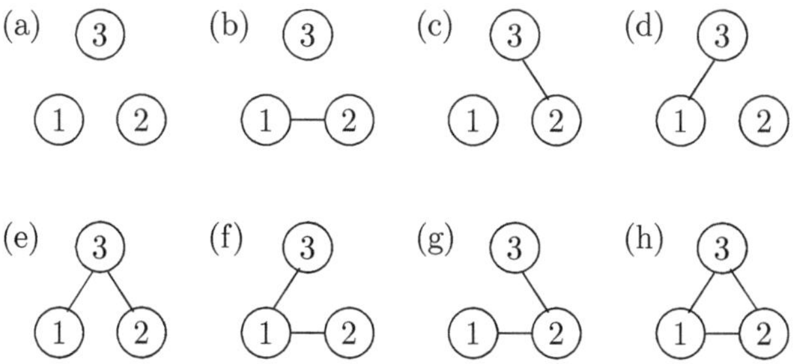

Fig. 3.2 For an undirected graph with three vertices, there are 2^3 possible types depending on the presence or absence of edges

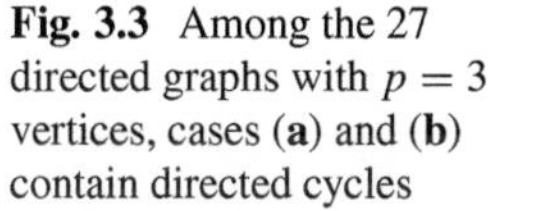

Fig. 3.3 Among the 27 directed graphs with $p = 3$ vertices, cases (**a**) and (**b**) contain directed cycles

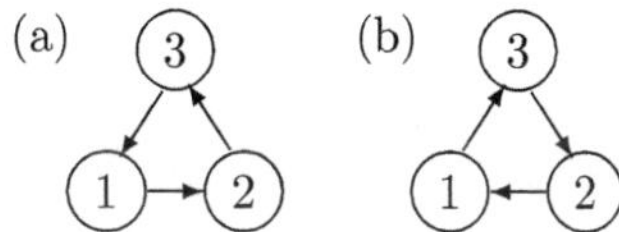

In a **directed graph**, a directed edge from vertex i to j is denoted by the ordered pair (i, j), so $(i, j) \neq (j, i)$. The edge set E is a subset of $\mathcal{E} := \{(i, j) \mid i, j \in V,\ i \neq j\}$. If $(i, j) \in E$, we say that there exists a **directed path** from i to j or that j is a **descendant** of i. If there exists a directed path from i to j and $(j, k) \in E$, then there exists a directed path from i to k. If all vertices on a directed path are distinct, it is called a **directed acyclic path**. If all directed paths are acyclic, the directed graph $G = (V, E)$ is called a **directed acyclic graph (DAG)**. Below, we will refer to directed acyclic graphs as DAGs. A path that ignores the direction of each edge and follows connections is called an **undirected path**.

Example 3.4 (DAG) As there are 2^3 undirected graphs when $V = \{1, 2, 3\}$, there are $3^3 = 27$ directed graphs on $p = 3$ vertices (for each pair of vertices, there are three possibilities: no edge, an edge in one direction, or in the other). However, two of these, shown in Fig. 3.3, contain directed cycles, so there are 25 DAGs. ■

Next we define separation in graphs.

First consider, in an undirected graph, the undirected paths (possibly multiple) from i to j for two distinct vertices $i, j \in V$. Let C be a subset of V that does not contain i or j. If every path from i to j passes through at least one vertex in C, we say that i and j are **separated** by C.

If A, B, C are pairwise disjoint subsets of V and every $i \in A$ and $j \in B$ is separated by C, then we say that A and B are separated by C and write $A \perp\!\!\!\perp_G B \mid C$.

Example 3.5 (Separation in an Undirected Graph) For the undirected graph in Fig. 3.5a, assess the truth of the following:

1. $\{5\} \perp\!\!\!\perp_G \{12\} \mid \{9\}$: There exists a path $5 \to 8 \to 12$ not meeting $\{9\}$; false.
2. $\{7\} \perp\!\!\!\perp_G \{9\}$: There exists a path $7 \to 3 \to 5 \to 9$; false.

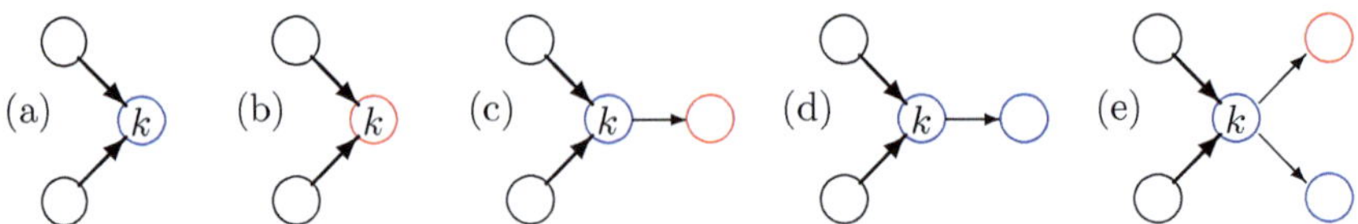

Fig. 3.4 (**a**) When a collision occurs at k on the path, it is separated since $k \notin C$ and has no descendants, (**b**) not separated because $k \in C$, (**c**) not separated because $k \notin C$ but a descendant is in C, (**d**) separated because $k \notin C$ and none of its descendants are in C, and (**e**) not separated because $k \notin C$ but a descendant is in C

3. $\{1\} \perp\!\!\!\perp_G \{10\} \mid \{5, 8\}$: Any path from 1 to 10 must pass through 5 or 8; true.
4. $\{7\} \perp\!\!\!\perp_G \{10\} \mid \{13\}$: Any path from 7 to 10 must pass through 13; true.
5. $\{1\} \perp\!\!\!\perp_G \{12\} \mid \{4, 9, 11\}$: There is a path $1 \to 2 \to 5 \to 8 \to 12$ not meeting $\{4, 9, 11\}$; false.
6. $\{2, 4\} \perp\!\!\!\perp_G \{6, 13\} \mid \{5, 12\}$: Every path connecting 2 to 6, 2 to 13, 4 to 6, and 4 to 13 must pass through 5 or 12; true.

■

Next consider DAGs. For two distinct vertices $i, j \in V$, consider the undirected paths (possibly multiple) between i and j. Let C be a subset of V not containing i or j. If every undirected path from i to j contains a vertex k satisfying one of the following conditions, then we say i and j are **d-separated** by C (see Fig. 3.4).

1. The directions of the two edges adjacent to k along the path are $\to k \to$, $\leftarrow k \to$, or $\leftarrow k \to$, and $k \in C$.
2. The directions are $\to k \leftarrow$ (a **collider**), and **neither** k **nor any of its descendants belongs to** C.

If A, B, C are pairwise disjoint subsets of V and every $i \in A$ and $j \in B$ is d-separated by C, then we say A and B are d-separated by C and write $A \perp\!\!\!\perp_G B \mid C$. In this book, the case "$\to k \leftarrow$" is called a **collider (head to head)**.

Example 3.6 (D-Separation in a DAG) From Fig. 3.5b, assess the truth of the following:

1. $\{5\} \perp\!\!\!\perp_G \{12\} \mid \{9\}$: Both $5 \to 9 \to 12$ and $5 \to 8 \to 12$ are blocked; true. Indeed, 8 and its descendant 11 are not in the separating set $\{9\}$, so although $5 \to 8 \leftarrow 12$ is a collider, it remains blocked.
2. $\{7\} \perp\!\!\!\perp_G \{9\}$: There exists a path $7 \to 3 \to 5 \to 9$ that is not blocked; false.

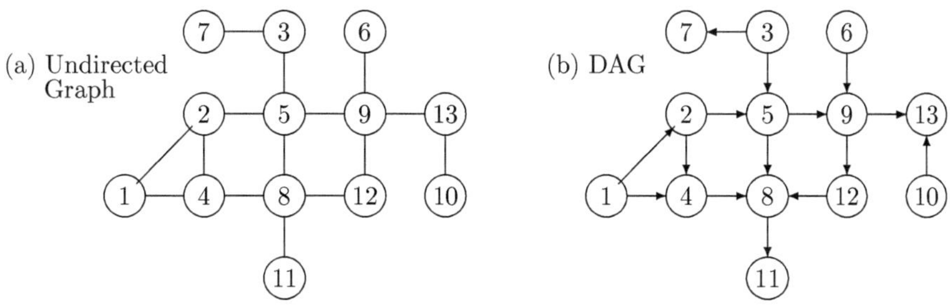

Fig. 3.5 The undirected graph in Example 3.5 and the DAG in Example 3.6

3. $\{1\} \perp\!\!\!\perp_G \{10\} \mid \{5, 8\}$: Every path goes through 13, which is a collider not included in the separating set, thus blocked; true.
4. $\{7\} \perp\!\!\!\perp_G \{10\} \mid \{13\}$: Any path from 7 to 10 must pass through 13, but 13 is a collider and is included in the separating set $\{13\}$, which opens the collider, hence not blocked; false.
5. $\{1\} \perp\!\!\!\perp_G \{12\} \mid \{4, 9, 11\}$: The path $1 \to 2 \to 5 \to 8 \to 12$ has a collider at 8, but its descendant 11 is in the separating set $\{4, 9, 11\}$, so the path is unblocked; false.
6. $\{2, 4\} \perp\!\!\!\perp_G \{6, 13\} \mid \{5, 12\}$: Every path connecting 2 to 6 or 13, and 4 to 6 or 13, must pass through 5 or 12; true.

■

Whether we use an undirected graph or a DAG will be clear from context, so we use the same notation $G = (V, E)$. Separation applies to undirected graphs and d-separation to DAGs.

The following proposition holds for both separation and d-separation.

Proposition 3.2 *Let A, B, C, D be pairwise disjoint subsets of $\{1, \ldots, p\}$. Then*

$$A \perp\!\!\!\perp_G C \mid D \text{ and } B \perp\!\!\!\perp_G C \mid D \iff (A \cup B) \perp\!\!\!\perp_G C \mid D, \tag{3.8}$$

Proof See Problem 25. ■

3.3 Representing Conditional Independence with Graphs

A **graphical model** is the representation of conditional independence among random variables via separation properties in a graph. We identify random variables with vertices.

Let A, B, C be disjoint subsets of the vertex set $V = \{1, 2, \dots, p\}$, identified with $\{X_1, \dots, X_p\}$, and choose the edge set E so that

$$A \perp\!\!\!\perp_G B \mid C \Longrightarrow X_A \perp\!\!\!\perp X_B \mid X_C. \tag{3.9}$$

However, the converse

$$A \perp\!\!\!\perp_G B \mid C \Longleftarrow X_A \perp\!\!\!\perp X_B \mid X_C \tag{3.10}$$

need not hold for any E.

If we start from $E = \mathcal{E}$ (all possible edges), then nothing is separated, and $A \perp\!\!\!\perp_G B \mid C$ is false, so (3.9) holds trivially. From there, we successively remove edges while maintaining (3.9) and minimize the number of edges (details in Example 3.9). When G is undirected, such a model is called a **Markov network** (MN), and when G is a DAG, it is called a **Bayesian network** (BN) [13, 19, 28]. In general, even when the set of conditional independencies to be represented is fixed, the edge set E need not be unique. We illustrate this with examples.

Example 3.7 Equation (3.2) in Example 3.2 implies $X \perp\!\!\!\perp Y \mid Z$, and the MN representation is shown in Fig. 3.1a, while BN representations are shown in Fig. 3.1b,c,d. In (a), $X \perp\!\!\!\perp_G Y \mid Z$ holds by (undirected) separation; in (b), (c), and (d), $X \perp\!\!\!\perp_G Y \mid Z$ holds by d-separation. There are no other nontrivial separations or d-separations in these graphs. Thus, in this case, (3.10) also holds. ■

As seen in Example 3.7, a BN is generally not unique. Figure 3.1b,c,d corresponds to the three factorizations (3.3) and represents the same distribution. BNs that represent the same distribution are said to be **Markov equivalent**.

Example 3.8 There are 25 DAGs on $p = 3$ vertices (Example 3.4), but identifying Markov equivalent BNs yields the 11 classes in Fig. 3.6 (see Problem 19). For example,

$$P(X)\,P(Y)\,P(Z),$$

$$P(X)\,P(YZ),\;\; P(Y)\,P(ZX),\;\; P(Z)\,P(XY),$$

$$\frac{P(ZX)\,P(XY)}{P(X)},\;\; \frac{P(XY)\,P(YZ)}{P(Y)},\;\; \frac{P(ZX)\,P(XY)}{P(Z)},$$

$$\frac{P(Y)\,P(Z)\,P(XYZ)}{P(YZ)},\;\; \frac{P(Z)\,P(X)\,P(XYZ)}{P(ZX)},\;\; \frac{P(X)\,P(Y)\,P(XYZ)}{P(XY)},$$

$$P(XYZ).$$

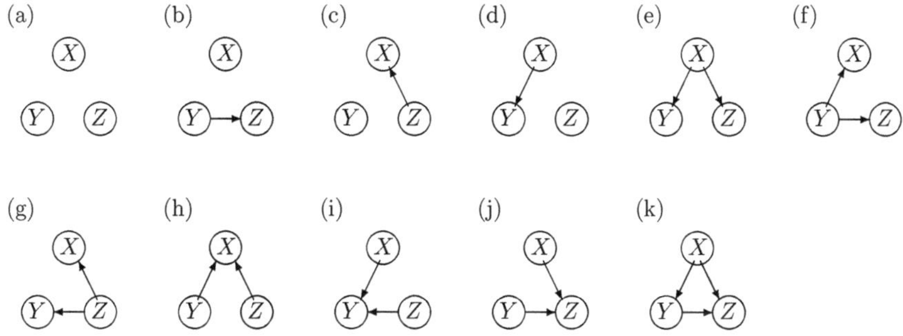

Fig. 3.6 The 11 Markov equivalent Bayesian networks for $p = 3$

Among these, in (a) one can simultaneously represent multiple conditional independencies such as

$$X \perp\!\!\!\perp Y, \quad Y \perp\!\!\!\perp Z, \quad Z \perp\!\!\!\perp X, \quad X \perp\!\!\!\perp \{Y, Z\}, \quad Y \perp\!\!\!\perp \{Z, X\}, \quad Z \perp\!\!\!\perp \{X, Y\},$$

while in (k) no nontrivial conditional independencies are represented. ■

For a given set of conditional independencies, sometimes a BN is suitable, and other times an MN is more appropriate.

Example 3.9 (When a BN Is Appropriate) In Fig. 3.6h,i,j, respectively,

$$Y \perp\!\!\!\perp_G Z, \qquad Z \perp\!\!\!\perp_G X, \qquad X \perp\!\!\!\perp_G Y$$

hold, and from the factorized forms we also have

$$Y \perp\!\!\!\perp Z, \qquad Z \perp\!\!\!\perp X, \qquad X \perp\!\!\!\perp Y.$$

Thus (3.10) holds in each case. However, an MN cannot represent these nontrivial (here, marginal) independencies. Indeed, there are eight MNs on $p = 3$ (Fig. 3.7). In (h), no nontrivial conditional independence is asserted, so (3.9) holds trivially. Removing any edge from (h) yields one of (e), (f), or (g), which assert, respectively,

$$Y \perp\!\!\!\perp Z \mid X, \qquad Z \perp\!\!\!\perp X \mid Y, \qquad X \perp\!\!\!\perp Y \mid Z,$$

and these do not follow from $Y \perp\!\!\!\perp Z$, etc. Hence no edges can be removed from (h). Therefore the MN representing these independencies is (h), which asserts no nontrivial conditional independencies. ■

Example 3.10 (When an MN Is Appropriate) Consider random variables X, Y, Z, W with joint distribution

$$P(X, Y, Z, W) = \frac{P(X, Z)\, P(Z, Y)\, P(Y, W)\, P(W, X)}{P(X)\, P(Y)\, P(Z)\, P(W)}.$$

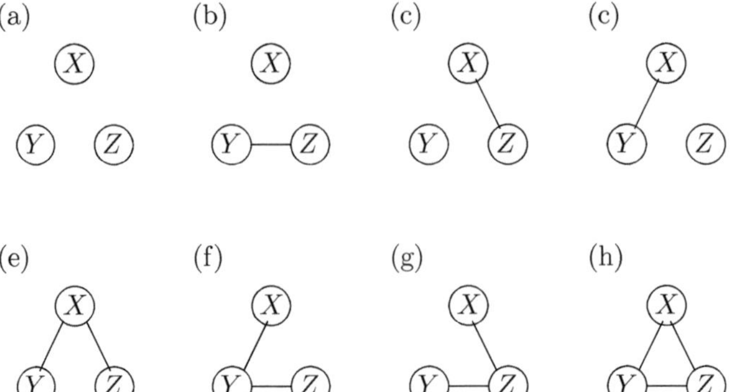

Fig. 3.7 The eight Markov networks for $p = 3$

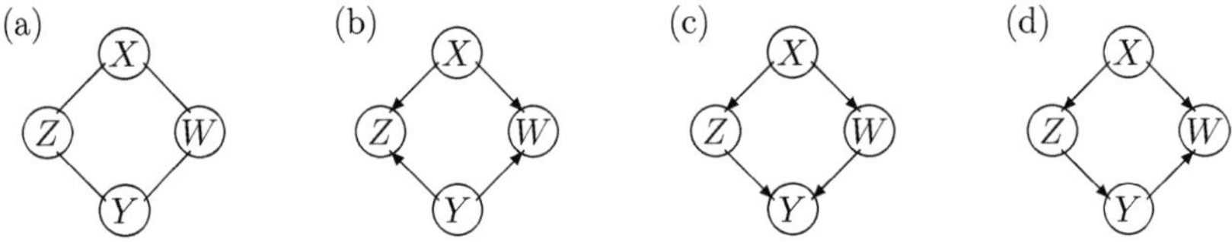

Fig. 3.8 In (**a**), both $X \perp\!\!\!\perp_G Y \mid \{Z, W\}$ and $Z \perp\!\!\!\perp_G W \mid \{X, Y\}$ hold, in (**c**), only $X \perp\!\!\!\perp_G Y \mid \{Z, W\}$ holds, and in (**b**) and (**d**), only $Z \perp\!\!\!\perp_G W \mid \{X, Y\}$ holds

Then the conditional independencies $X \perp\!\!\!\perp Y \mid \{Z, W\}$ and $Z \perp\!\!\!\perp W \mid \{X, Y\}$ hold. Indeed,

$$P(X, Y, Z, W) = \frac{1}{P(Z)P(W)} \cdot \frac{P(W, X)\, P(X, Z)}{P(X)} \cdot \frac{P(W, Y)\, P(Y, Z)}{P(Y)},$$

and marginalizing over X, Y, and (X, Y) yields, respectively,

$$\begin{cases} P(Y, Z, W) = \dfrac{P(W, Y)\, P(Y, Z)}{P(Y)}, \\ P(X, Z, W) = \dfrac{P(W, X)\, P(X, Z)}{P(X)}, \end{cases} \qquad P(Z, W) = P(Z)\, P(W). \tag{3.11}$$

Therefore,

$$P(X, Y \mid Z, W) = P(X \mid Z, W)\, P(Y \mid Z, W) \tag{3.12}$$

holds. A similar derivation gives $Z \perp\!\!\!\perp W \mid \{X, Y\}$ (see Problem 26). In the MN of Fig. 3.8a, we have $X \perp\!\!\!\perp_G Y \mid \{Z, W\}$ and $Z \perp\!\!\!\perp_G W \mid \{X, Y\}$, so both (3.9) and (3.10) hold. However, attempting the same with a BN while avoiding cycles yields Fig. 3.8b, where only $Z \perp\!\!\!\perp_G W \mid \{X, Y\}$ holds, and Figs. 3.8c,d, where only $X \perp\!\!\!\perp_G Y \mid \{Z, W\}$ holds. Removing any further edges would assert independencies that do not hold. Hence there is no BN for which both (3.9) and (3.10) hold simultaneously. ■

In later chapters, when both (3.9) and (3.10) hold for a BN, we say that the set of random variables is **faithful** to that BN, meaning that all and only the conditional independencies of the distribution are represented by the BN.

3.4 The Chow–Liu Algorithm

Consider an undirected graph G with vertex set $V = \{1, \ldots, p\}$. For each unordered pair $i \neq j$, assign a weight $w_{i,j}$, and seek an edge set E maximizing the sum of weights. Assume the weights are positive and symmetric ($w_{i,j} = w_{j,i}$).

In particular, suppose $G = (V, E)$ is a **spanning tree**, i.e., a connected acyclic undirected graph. It is known that the following procedure [12] finds an optimal G (note that if multiple pairs share the same weight, the optimal tree need not be unique).

Kruskal Algorithm Initialize $\mathcal{E} = \{\{i, j\} \mid i \neq j\}$ and $E = \emptyset$. Repeat until $\mathcal{E} = \emptyset$:

1. Remove from $\mathcal{E}$ an edge $\{i, j\}$ with the largest weight $w_{i,j}$ (equivalently, process edges in descending order of weight).
2. If adding $\{i, j\}$ to E does not create a cycle, add it to E.

Example 3.11 For $V = \{1, \ldots, 4\}$ with weights satisfying $w_{1,2} > w_{2,3} > w_{1,3} > w_{1,4} > w_{2,4} > w_{3,4} > 0$ (all other entries zero), the forest evolves in the order shown in Fig. 3.9. ■

Proposition 3.3 *Kruskal algorithm yields a spanning tree whose total edge weight is maximized.*

Proof See [12] or the Wikipedia article "Kruskal Algorithm." Kruskal algorithm can be adapted as follows without changing its behavior:

- Negative edge weights are allowed.
- One may obtain a maximum-weight forest instead of a spanning tree.

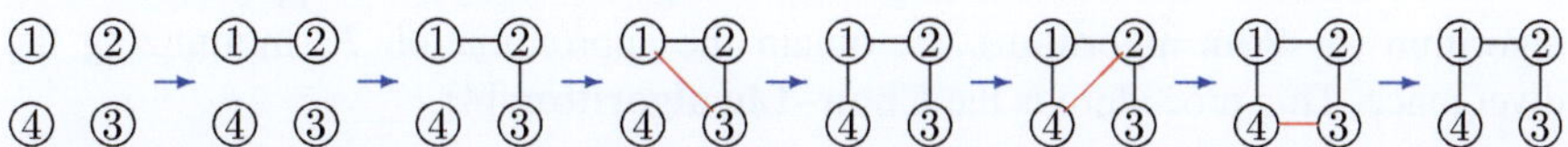

Fig. 3.9 Following the order of weights, edges $\{1, 2\}$, $\{2, 3\}$, $\{1, 3\}$, $\{1, 4\}$, $\{2, 4\}$, and $\{3, 4\}$ are added in this sequence. However, when attempting to add $\{1, 3\}$, a cycle is created, so $\{1, 3\}$ is not included. The same applies to $\{2, 4\}$ and $\{3, 4\}$

Since the goal is to maximize the total weight, vertices connected only by negative-weight edges will remain disconnected, yielding a forest rather than a tree. ■

In practice, **Prim algorithm** [20] is often preferred for efficiency.[1] See the Appendix for details.

Next, for discrete random variables $X_1, \ldots, X_p$ with joint distribution

$$P(X_1 = x_1, \ldots, X_p = x_p), \qquad x_k \in \mathcal{X}_k,$$

where the sets of the values $X_1, \ldots, X_p$ take are denoted by $\mathcal{X}_1, \ldots, \mathcal{X}_p$, we consider approximating this distribution by one representable by a tree. Specifically, with $V = \{1, \ldots, p\}$ and $E \subset \{\{i, j\} \mid i \neq j\}$, approximate by

$$\prod_{k \in V} P(X_k = x_k) \prod_{\{i,j\} \in E} \frac{P(X_i = x_i, X_j = x_j)}{P(X_i = x_i)\, P(X_j = x_j)},$$

where E is chosen to be acyclic (a forest). For brevity, write $P(x_1, \ldots, x_p)$ for the original probability and $P'(x_1, \ldots, x_p)$ for the approximation. The Kullback–Leibler divergence between them is

$$D(P \| P') := \sum_{x_1} \cdots \sum_{x_p} P(x_1, \ldots, x_p) \log \frac{P(x_1, \ldots, x_p)}{P'(x_1, \ldots, x_p)}. \tag{3.13}$$

Define the **mutual information** between X_i and X_j as

$$I(i, j) := \sum_{x_i} \sum_{x_j} P(x_i, x_j) \log \frac{P(x_i, x_j)}{P(x_i)\, P(x_j)}.$$

Let $H(k)$ denote the entropy of X_k, and $H(1, \ldots, p)$ the joint entropy. Then (see Problem 29)

$$D(P \| P') = -H(1, \ldots, p) + \sum_{k \in V} H(k) - \sum_{\{i,j\} \in E} I(i, j). \tag{3.14}$$

The first two terms on the right-hand side of (3.14) do not depend on E, so minimizing $D(P \| P')$ amounts to maximizing the sum of mutual information in the third term. Therefore, by taking weights $w_{i,j} = I(i, j)$ and applying Kruskal algorithm (or Prim algorithm), we obtain the approximation P' minimizing the divergence. This procedure is the **Chow–Liu algorithm** [4].

[1] Historically, the algorithm was first published by Vojtěch Jarník in 1930; Prim re-published it in 1957.

Appendix

Prim Method

Given a subset $U \subset V$, at each step include in U a vertex $i \in V \setminus U$ maximizing $w[i, j]$ for some $j \in U$, and add the edge $\{i, j\}$.

Prim Algorithm Initialize $U = \emptyset$, $v[i] = 0$ $(i \in V)$, and $j \leftarrow 1$.

1. Add j to U. For each $i \notin U$ with $w[i, j] > v[i]$, set $v[i] \leftarrow w[i, j]$ and $\pi[i] \leftarrow j$.
2. Set j to be an $i \notin U$ maximizing $v[i]$.

Finally, the edges $\{i, \pi[i]\}$, $i = 2, \ldots, p$, form a maximum spanning tree.

Viewing a tree as rooted at some vertex, we may interpret edges as directed away from the root; The head i is the child and the tail $\pi[i]$ the parent. Any vertex can serve as the root (above, $j = 1$), and every non-root vertex has exactly one parent. Step 1 updates $v[i] \leftarrow \max_{j \in U} w[i, j]$ for each $i \notin U$ (and records $\pi[i]$ when updated). Prim method constructs, for each $1 \leq k \leq p$, a maximum-weight tree on a subset U of size k containing 1. This holds trivially for $|U| = 1$ $(U = \{1\})$. Assuming it holds for $|U| = k$, Step 2 adds the edge with weight

$$\max_{i \notin U} \max_{j \in U} w[i, j] = \max_{i \notin U} w[i, \pi[i]] = \max_{i \notin U} v[i],$$

so the total weight remains maximal when $|U| = k + 1$ (each $i \notin U$ stores $\pi[i]$ attaining its current maximum). The claim also holds for $k = p$.

The above assumes a tree; for a forest, we look for isolated vertices and start a new set U.

The following is a Python implementation (using the maximum total weight). Note it assumes a (possibly) disconnected graph and thus returns a maximum forest.

```
import numpy as np
```

```
def kruskal(w):
    w = w.copy()
    p = w.shape[1]
    for i in range(p - 1):
        for j in range(i + 1, p):
            w[j, i] = w[i, j]
    for i in range(p):
        w[i, i] = 0
    parent = np.empty(p, dtype=int)
    u = np.zeros(p, dtype=int)
    v = np.full(p, -np.inf, dtype=float)
    j = 0
    parent[0] = -1
    while j <= p - 1:
        u[j] = 1
        for i in range(p):
```

```
            if u[i] == 0 and w[j, i] > v[i]:
                v[i] = w[j, i]
                parent[i] = j
        max_weight = 0.0
        for i in range(p):
            if u[i] == 0 and v[i] > max_weight:
                j = i
                max_weight = v[i]
        if max_weight <= 0:
            j = 0
            while j <= p - 1 and u[j] == 1:
                j += 1
            if j <= p - 1:
                parent[j] = -1
        if j > p - 1:
            break
    pair_1 = []
    pair_2 = []
    for j in range(p):
        if parent[j] != -1:
            pair_1.append(parent[j])
            pair_2.append(j)
    return np.column_stack((pair_1, pair_2))
```

In this code, multiple connected components (a forest) are allowed. If a spanning tree is guaranteed, the block `if max_weight <= 0:` can be removed.

Example 3.12 Using `kruskal`, we compute the edge list (`edgelist`) and visualize it with the `igraph` package. Example outputs are shown in Fig. 3.10.

```
import networkx as nx
import matplotlib.pyplot as plt
p = 10
w = np.random.randn(p * p).reshape(p, p) ** 2
edgelist = kruskal(w)
G = nx.Graph()
G.add_edges_from([tuple(map(int, e)) for e in edgelist.tolist()])
nx.draw(G, with_labels=True)
plt.show()
```

■

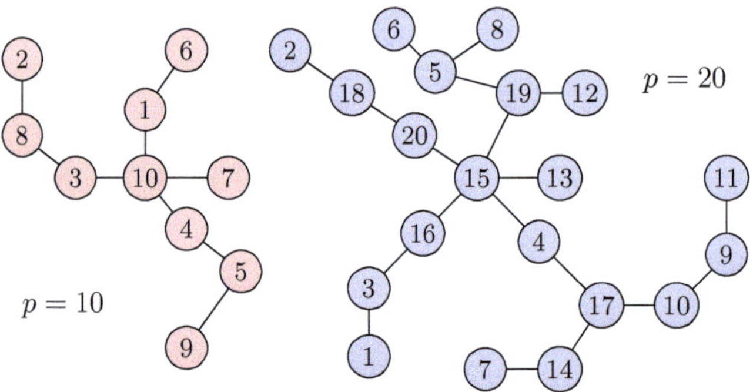

Fig. 3.10 Graphs obtained by running the program in Example 3.12. Left: $p = 10$; right: $p = 20$. Both are spanning trees

Problems 19–29

19. For each of the 11 Bayesian networks (BNs) in Fig. 3.6, identify which of the 25 directed acyclic graphs (DAGs) they represent.
20. The distributions of the 11 BNs in Fig. 3.6 are given in Example 3.8. What are the distributions of the eight Markov networks (MNs) in Fig. 3.7?
21. When a collider occurs at k on a path, state whether k d-separates in each of the following four cases:

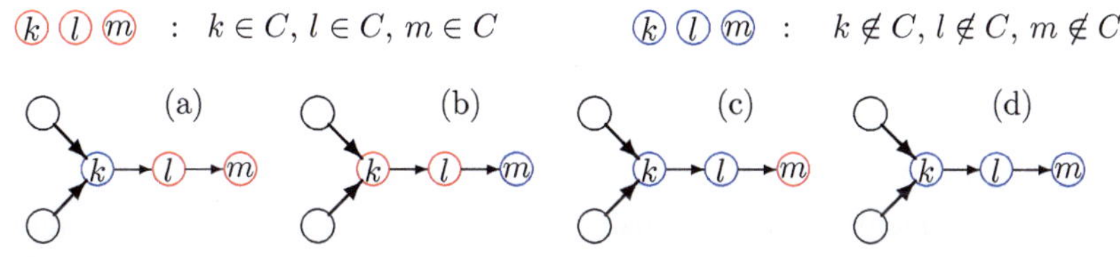

22. For the undirected graph in Fig. 3.5a, determine whether each of the following propositions is true or false:

    (a) $3 \perp\!\!\!\perp_G 9 \mid \{5, 8\}$
    (b) $7 \perp\!\!\!\perp_G 12 \mid \{3, 8\}$
    (c) $9 \perp\!\!\!\perp_G 12 \mid 8$
    (d) $\{1, 2\} \perp\!\!\!\perp_G 5 \mid 3$
    (e) $\{4, 5\} \perp\!\!\!\perp_G 6 \mid 7$
    (f) $\{7, 9\} \perp\!\!\!\perp_G 10 \mid 8$
    (g) $\{11, 9\} \perp\!\!\!\perp_G 12 \mid 13$
    (h) $\{1, 13\} \perp\!\!\!\perp_G 9 \mid 4$

23. For the DAG in Fig. 3.5b, determine whether each of the following propositions is true or false:

    (a) $3 \perp\!\!\!\perp_G 10 \mid \{5, 8\}$
    (b) $7 \perp\!\!\!\perp_G 12 \mid \{3, 8\}$
    (c) $5 \perp\!\!\!\perp_G 12 \mid 9$
    (d) $1 \perp\!\!\!\perp_G 12 \mid \{4, 8\}$
    (e) $2 \perp\!\!\!\perp_G 9 \mid \{5, 8\}$
    (f) $3 \perp\!\!\!\perp_G 10 \mid 5$
    (g) $7 \perp\!\!\!\perp_G 12 \mid \{3, 9\}$
    (h) $11 \perp\!\!\!\perp_G 9 \mid 10$

24. Show that even if edges are added to Fig. 3.8b,c,d in Example 3.10, the same separation relations as in Fig. 3.8a cannot be obtained. Further, show that if edges are deleted from (b),(c),(d), this would imply false conditional independencies.
25. Prove Proposition 3.2.
26. Explain why marginalization yields (3.11), and then use (3.11) to show (3.12).

27. Let X and Y be independent random variables taking values ± 1, and define $Z = XY$. Determine the Markov network (MN) and Bayesian network (BN) of X, Y, Z.
28. In the function `kruskal`:

    (a) When the input matrix w is not symmetric, after the initial symmetrization (copying the upper triangle to the lower triangle) and then zeroing the diagonal, what kind of matrix does w become?
    (b) What is the relationship between whether a vertex j belongs to the set U and the value of u[j]?
    (c) In the step where a new root for a connected component is chosen from unvisited vertices, for which value of u[j] is parent[j] set to -1? Also, how is vertex j treated at that time?

29. (a) Show that (3.13) takes nonnegative values and equals 0 only when P and P' coincide. Also, state under what conditions (3.13) takes finite values.
    (b) Show that the mutual information $I(i, j)$ corresponds to a Kullback–Leibler divergence and hence is nonnegative. Also, state when the mutual information equals 0.
    (c) Show equality (3.14).

Chapter 4
Testing Independence and Conditional Independence with Kernels

In this chapter, we explain methods for testing independence and conditional independence between variables using kernel functions. Kernel methods are powerful techniques that flexibly handle nonlinear and high-dimensional dependence structures.

We begin with an intuitive introduction to the basic concepts of reproducing kernel Hilbert spaces (RKHSs). We then present the Hilbert–Schmidt Independence Criterion (HSIC) for testing the independence of two variables, along with its theoretical background and implementation. Next, we turn to Kernel Conditional Independence (KCI), which tests independence conditional on a third variable.

These methods construct test statistics by exploiting covariance structures in the feature spaces defined by kernels. We also discuss estimation procedures and implementation caveats for real data applications.

The kernel-based tests introduced in this chapter play a central role in estimating causal structures in Chaps. 5 and 6.

4.1 Reproducing Kernel Hilbert Spaces

Let $\mathcal{X}$ be a set. A mapping $k : \mathcal{X} \times \mathcal{X} \to \mathbb{R}$ is called a **positive definite kernel** if it satisfies the following two conditions.

1. Symmetry: $k(x, y) = k(y, x)$ for $x, y \in \mathcal{X}$.
2. Positive definiteness: For any $N \geq 1$ and $x_1, \ldots, x_N \in \mathcal{X}$, the matrix $K \in \mathbb{R}^{N \times N}$ with entries $k(x_i, x_j)$ is positive semidefinite.

In particular, the matrix K above is called the **Gram matrix**. Note that positive definiteness here requires the Gram matrix to be positive semidefinite rather than strictly positive definite.

J. Suzuki, *Graphical Models and Causal Discovery with Python*,
https://doi.org/10.1007/978-981-95-5308-2_4

In general, the term "kernel" is often used to refer to a measure of similarity within $\mathcal{X}$, and kernels that do not satisfy positive definiteness are also used in practice.

Example 4.1 (Epanechnikov Kernel) Let $\lambda > 0$. For $x, y \in \mathbb{R}$, define

$$k(x, y) = D\left(\frac{|x - y|}{\lambda}\right),$$

$$D(t) = \begin{cases} \frac{3}{4}(1 - t^2), & |t| \leq 1, \\ 0, & \text{otherwise.} \end{cases}$$

Apply this kernel to the smoothing method known as the **Nadaraya–Watson estimator**. Let observed data $(x_1, y_1), \ldots, (x_N, y_N) \in \mathcal{X} \times \mathbb{R}$ be given, and suppose a new $x_* \in \mathcal{X}$ is provided. Then return as $\hat{f}(x_*)$ the weighted average of $y_1, \ldots, y_N$ with weights

$$\frac{k(x_*, x_1)}{\sum_{j=1}^{N} k(x_*, x_j)}, \ldots, \frac{k(x_*, x_N)}{\sum_{j=1}^{N} k(x_*, x_j)}.$$

Since $k(x, y)$ is assumed to be large when $x, y \in \mathcal{X}$ are similar, larger weight is assigned to y_i whose x_i is more similar to x_*. As λ decreases, the prediction uses only (x_i, y_i) in a neighborhood of x_*. Figure 4.1 shows the result of this procedure. ∎

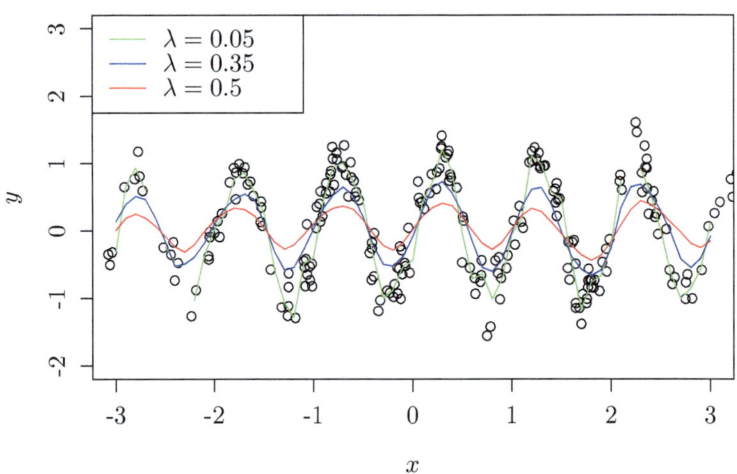

Fig. 4.1 Applying the Epanechnikov kernel to the Nadaraya–Watson estimator; curves are drawn for $\lambda = 0.05, 0.35, 0.5$

The kernel in Example 4.1 is not positive definite. For instance, with $\lambda = 2$, $n = 3$, and $x_1 = -1,\ x_2 = 0,\ x_3 = 1$,

$$K_\lambda = \begin{bmatrix} 3/4 & 9/16 & 0 \\ 9/16 & 3/4 & 9/16 \\ 0 & 9/16 & 3/4 \end{bmatrix}$$

one of the eigenvalues is negative (Problem 31).

On the other hand, the Gaussian kernel

$$k(x, y) = \exp\left\{-\frac{1}{2\sigma^2}\|x - y\|^2\right\}, \qquad \sigma^2 > 0,$$

and the Cauchy kernel

$$k(x, y) = \frac{1}{2\pi} \cdot \frac{1}{\|x - y\|^2 + \beta^2}, \qquad \beta > 0,$$

are positive definite when $\mathcal{X} = \mathbb{R}^p$. Here, for $a = [a_1, \ldots, a_p]$, set $\|a\| = \sqrt{\sum_{j=1}^p a_j^2}$. Positive definiteness follows from the fact that these $k(x, y)$ depend only on $\|x - y\|$ and are constant multiples of characteristic functions of probability densities,

$$f_X(x) = \frac{1}{\sqrt{2\pi\sigma^2}} \exp\left\{-\frac{x^2}{2\sigma^2}\right\}, \qquad f_X(x) = \frac{\beta}{2} \exp\{-\beta|x|\},$$

namely

$$\phi(t) = \int_{\mathcal{X}} e^{it^\top x} f_X(x)\, dx$$

(Problem 32). Indeed, the following proposition holds.

Proposition 4.1 (Bochner's Theorem [2]) *For any $N \geq 1$, $x_1, \ldots, x_N \in \mathcal{X}$, and $z_1, \ldots, z_N \in \mathbb{R}$,*

$$\sum_{i=1}^N \sum_{j=1}^N z_i z_j \phi(x_i - x_j) \geq 0$$

holds if and only if $\phi(\cdot)$ is a constant multiple of the characteristic function of some probability density.[1]

[1] A proof can be found in J. Suzuki, *100 Kernel Problems for Machine Learning with R/Python* [33, 34].

In addition, the inner product of a finite-dimensional Euclidean space also satisfies the definition of a kernel (Problem 33).

We will discuss later why positive definiteness is required for kernels; first, we record some properties.

For each $x \in \mathcal{X}$, consider the function

$$\mathcal{X} \ni y \mapsto k(x, y) \in \mathbb{R}.$$

Let H_0 be the set of functions f expressible as $f = \sum_{i=1}^{N} \alpha_i k(x_i, \cdot)$ for some $N \geq 1, x_1, \ldots, x_N \in \mathcal{X}$, and $\alpha_1, \ldots, \alpha_N \in \mathbb{R}$. Then H_0 is a **linear space**. That is, for any $\alpha \in \mathbb{R}$ and $f, g \in H_0$,

$$\alpha f \in H_0, \qquad f + g \in H_0 \tag{4.1}$$

(Problem 34). Define the **inner product** on H_0 by $\langle f, g \rangle = \sum_{i=1}^{N} \sum_{j=1}^{N} \alpha_i \beta_j k(x_i, x_j)$ for $f = \sum_{i=1}^{N} \alpha_i k(x_i, \cdot)$ and $g = \sum_{j=1}^{N} \beta_j k(x_j, \cdot)$. Then the four axioms of inner products hold for all $\alpha, \beta \in \mathbb{R}$ and $f, g, h \in H_0$:

$$\langle f, f \rangle \geq 0, \quad \langle \alpha f + \beta g, h \rangle = \alpha \langle f, h \rangle + \beta \langle g, h \rangle, \quad \langle f, g \rangle = \langle g, f \rangle,$$
$$\langle f, f \rangle = 0 \Longrightarrow f = 0 \tag{4.2}$$

(Problem 34). In this book we treat general linear spaces and inner products satisfying (4.1) and (4.2), rather than special cases such as the following example.

Example 4.2 (Two-Dimensional Euclidean Space) For $a_1, a_2, b_1, b_2 \in \mathbb{R}$ and $\alpha \in \mathbb{R}$,

$$\begin{bmatrix} a_1 \\ a_2 \end{bmatrix} + \begin{bmatrix} b_1 \\ b_2 \end{bmatrix} = \begin{bmatrix} a_1 + b_1 \\ a_2 + b_2 \end{bmatrix}, \qquad \alpha \begin{bmatrix} a_1 \\ a_2 \end{bmatrix} = \begin{bmatrix} \alpha a_1 \\ \alpha a_2 \end{bmatrix}.$$

This set of vectors forms a linear space, and the inner product of $\begin{bmatrix} a_1 \\ a_2 \end{bmatrix}$ and $\begin{bmatrix} b_1 \\ b_2 \end{bmatrix}$ is $a_1 b_1 + a_2 b_2$. ■

Next we define completion. The limit of a sequence of rational numbers $\{a_n\}$ need not be rational. However, even in such cases, for sufficiently large m, n we have $|a_m - a_n| \to 0$.

Example 4.3 Consider $a_1 = 1$ and $a_{n+1} = \dfrac{1}{2} a_n + \dfrac{1}{a_n}$. Its limit is $\sqrt{2} \notin \mathbb{Q}$. Moreover,

$$|a_m - a_n| \leq |a_m - \sqrt{2}| + |a_n - \sqrt{2}| \to 0.$$

■

A sequence for which $|a_m - a_n| \to 0$ as m, n become large is called a **Cauchy sequence**. Without proof, any Cauchy sequence of real numbers is known to converge in $\mathbb{R}$.

In a set equipped with a distance (a **metric space**), if every Cauchy sequence converges within the set with respect to that distance, the space is called **complete**. In Example 4.3, the absolute value defines the distance on $\mathbb{R}$. The rationals $\mathbb{Q}$ are dense in $\mathbb{R}$, that is, for any $a \in \mathbb{R}$ and $\epsilon > 0$, the interval $(a - \epsilon, a + \epsilon)$ contains a rational number. It is known that $\mathbb{R}$ is obtained by adding points to $\mathbb{Q}$ to make it complete.

A linear space H equipped with an inner product is called an **inner-product space**. The norm induced by the inner product $\langle \cdot, \cdot \rangle_H$ is $\|a\|_H = \sqrt{\langle a, a \rangle_H}$ for $a \in H$.

Proposition 4.2 *Let H_1 be an inner-product space with the distance induced by its inner product. Then there exists a complete inner-product space H_2 containing H_1 as a dense subspace, and such an H_2 is unique up to isomorphism.*[2,3]

The operation that obtains H_2 from H_1 in Proposition 4.2 is called **completion**. A complete inner-product space is called a **Hilbert space**. The space H_2 consists of the limits of Cauchy sequences in H_1.

Example 4.4 (L^2 Space) Let L^2 denote the set of functions $f : [-\pi, \pi] \to \mathbb{R}$ satisfying

$$\int_{-\pi}^{\pi} \{f(x)\}^2 \, dx < \infty.$$

Then L^2 is a linear space (Problem 35), and with the inner product

$$\langle f, g \rangle = \int_{-\pi}^{\pi} f(x) g(x) \, dx,$$

it is complete. That is, for any sequence $\{f_n\} \subset L^2$ such that

$$\|f_m - f_n\| = \sqrt{\int_{-\pi}^{\pi} \{f_m(x) - f_n(x)\}^2 \, dx} \to 0 \quad (m, n \to \infty),$$

there exists $f \in L^2$ with $f_n \to f$. ■

Example 4.5 (Fourier Series Expansion) For a periodic function $f(x)$, we seek coefficients $a_0, a_1, \ldots$ and $b_1, b_2, \ldots$ such that

$$f(x) = a_0 + \sum_{n=1}^{\infty} (a_n \cos nx + b_n \sin nx).$$

---

[2] Two inner-product spaces are isomorphic if there exists a bijective linear isometry between them.

[3] For a proof, see, e.g., https://bayesnet.org/books_jp/?p=1656.

For example, define

$$f(x) = \begin{cases} 1, & 0 \le x < \pi, \\ 0, & \pi \le x < 2\pi, \end{cases}$$

and extend periodically by $f(x + 2\pi) = f(x)$. Then

$$f(x) = \frac{1}{2} + \frac{2}{\pi}\Big(\sin x + \frac{1}{3}\sin 3x + \frac{1}{5}\sin 5x + \cdots\Big).$$

Let

$$f_N(x) = \frac{1}{2} + \frac{2}{\pi}\Big(\sin x + \frac{1}{3}\sin 3x + \cdots + \frac{1}{2N-1}\sin\{(2N-1)x\}\Big).$$

Figure 4.2 shows the graphs for $N = 1, 2, 3, 4, 5, 6$. For any finite N we do not obtain $f(x)$ exactly. Moreover, $f(x)$ cannot be expressed as a finite linear combination of $\cos nx$ and $\sin nx$ $(n = 1, 2, \ldots)$. ■

For the inner-product space H_0 introduced in this section, there exists a Hilbert space H obtained by completing H_0. In particular, a Hilbert space obtained from a positive definite kernel is called a **reproducing kernel Hilbert space** (RKHS). The elements of H consist of limits of Cauchy sequences $\{f_n\}$ in H_0. For $f, g \in H$,

$$\langle f, g\rangle_H = \big\langle \lim_{N\to\infty} f_N, \lim_{N\to\infty} g_N\big\rangle_H := \lim_{N\to\infty} \langle f_N, g_N\rangle,$$

where $\langle\cdot,\cdot\rangle_H$ denotes the inner product in H. Even functions that are not finite linear combinations of $k(x,\cdot)$ can be represented as limits and belong to H. Linearity is preserved under taking limits, and the inner product is defined as the limit of inner products in H_0.

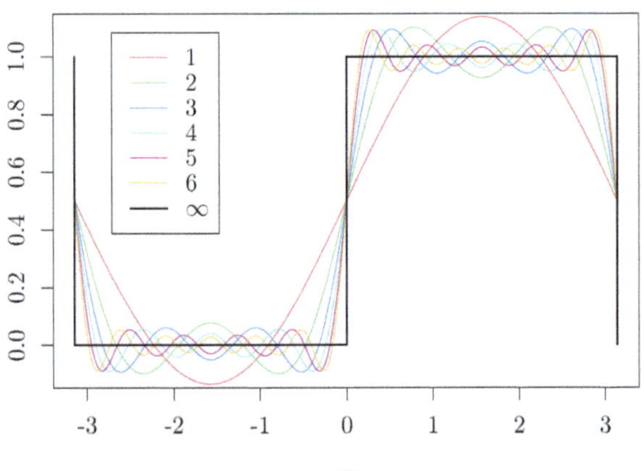

Fig. 4.2 Sketches of $f(x)$ $(N = \infty)$ and $f_N(x)$ for $N = 1, \ldots, 6$

An RKHS H and a positive definite kernel k are in one-to-one correspondence, and the following identity—known as the **kernel trick**—holds:

$$\langle f, k(x,\cdot)\rangle_H = f(x), \qquad x \in \mathcal{X},\ f \in H. \tag{4.3}$$

Since $f \in H$ is given as $\lim_{N\to\infty} f_N$ for some Cauchy sequence $\{f_N\} \subset H_0$, we obtain (4.3) as follows: Writing $f_N(\cdot) = \sum_{i=1}^N \alpha_i k(x_i,\cdot)$,

$$\begin{aligned}\langle f, k(x,\cdot)\rangle_H &= \left\langle \lim_{N\to\infty} f_N,\ k(x,\cdot)\right\rangle_H = \lim_{N\to\infty}\langle f_N, k(x,\cdot)\rangle_{H_0} \\ &= \lim_{N\to\infty}\left\langle \sum_{i=1}^N \alpha_i k(x_i,\cdot),\ k(x,\cdot)\right\rangle_{H_0} = \lim_{N\to\infty}\sum_{i=1}^N \alpha_i k(x_i,x) \\ &= \lim_{N\to\infty} f_N(x) = f(x).\end{aligned}$$

In data science and machine learning, kernels are used via the **feature map**

$$\mathcal{X} \ni x \mapsto k(x,\cdot) \in H,$$

which induces nonlinear representations. Concrete examples are given in Sects. 4.2 and 4.3 of this chapter.

4.2 Testing Independence: HSIC

Let $\mathcal{X}$ and $\mathcal{Y}$ be sets. Let $k_{\mathcal{X}} : \mathcal{X}\times\mathcal{X} \to \mathbb{R}$ and $k_{\mathcal{Y}} : \mathcal{Y}\times\mathcal{Y} \to \mathbb{R}$ be kernels, and let $H_{\mathcal{X}}$ and $H_{\mathcal{Y}}$ denote the corresponding RKHSs. In this section, we reduce the independence of $X \in \mathcal{X}$ and $Y \in \mathcal{Y}$ to the independence of $k_{\mathcal{X}}(X,\cdot) \in H_{\mathcal{X}}$ and $k_{\mathcal{Y}}(Y,\cdot) \in H_{\mathcal{Y}}$. Although $cov(X,Y) = 0$ does not imply independence of X and Y, here we define an analog of covariance for $k_{\mathcal{X}}(X,\cdot)$ and $k_{\mathcal{Y}}(Y,\cdot)$ and test whether it is zero.

In general, the product of two kernels is again positive definite. This follows from the proposition below.

Proposition 4.3 (Schur Product Theorem) *If $A = (a_{ij})$ and $B = (b_{ij})$ are positive semidefinite, then the* ***Hadamard product*** *$C = (a_{ij}b_{ij})$ is also positive semidefinite.*

Thus the RKHS $H_{\mathcal{X}\mathcal{Y}}$ corresponding to the positive definite kernel $k_{\mathcal{X}}k_{\mathcal{Y}}$ exists. By definition,

$$k_{\mathcal{X}\mathcal{Y}}(x,y,\cdot,\star) = k_{\mathcal{X}}(x,\cdot)\,k_{\mathcal{Y}}(y,\star), \qquad x \in \mathcal{X},\ y \in \mathcal{Y},$$

is an element of $H_{\mathcal{X}\mathcal{Y}}$[4]. Define the means

$$m_X := \mathbb{E}_X[k_{\mathcal{X}}(X,\cdot)], \quad m_Y := \mathbb{E}_Y[k_{\mathcal{Y}}(Y,\star)], \quad m_{XY} := \mathbb{E}_{XY}[k_{\mathcal{X}\mathcal{Y}}(X,Y,\cdot,\star)],$$

which belong to $H_{\mathcal{X}}$, $H_{\mathcal{Y}}$, and $H_{\mathcal{X}\mathcal{Y}}$, respectively. Since products of elements in $H_{\mathcal{X}}$ and $H_{\mathcal{Y}}$ lie in $H_{\mathcal{X}\mathcal{Y}}$, we may regard $m_X m_Y \in H_{\mathcal{X}\mathcal{Y}}$. In particular, we call $m_{XY} - m_X m_Y \in H_{\mathcal{X}\mathcal{Y}}$ the **covariance** of X and Y in the RKHS sense. Since $H_{\mathcal{X}\mathcal{Y}}$ is an RKHS, we can define

$$HSIC(X,Y) = \|m_{XY} - m_X m_Y\| \tag{4.4}$$

(the **Hilbert–Schmidt Independence Criterion**, HSIC). Gretton et al. [8] proposed using an estimator of this quantity to test independence.

Proposition 4.4 *If $k_{\mathcal{X}}$ and $k_{\mathcal{Y}}$ are both characteristic kernels, then*

$$X \perp\!\!\!\perp Y \Longleftrightarrow HSIC(X,Y) = 0.$$

A **characteristic kernel** is one for which the mapping

$$\mathcal{P} \ni P \mapsto m_X = \int_{\mathcal{X}} k(x,\cdot)\, dP(x)$$

is injective on a family $\mathcal{P}$ of probability distributions for X (if a density exists, write $dP(x) = f_X(x)dx$). In particular, when $k(x, y)$ is a function of $\|x - y\|$, the following is known.

Proposition 4.5 *A kernel $k : \mathcal{X}\times\mathcal{X} \to \mathbb{R}$ is characteristic if and only if the support of the corresponding probability density (whose characteristic function is k) is $\mathcal{X}$.*

Here, the support of f_X is the closure of $\{x \in \mathcal{X} : f_X(x) \neq 0\}$. Gaussian and Laplace kernels correspond to the characteristic functions of the normal and Laplace distributions, whose supports are all of $\mathcal{X} = \mathbb{R}$; hence they are characteristic.

Since the joint distribution of (X, Y) is unknown, $HSIC(X, Y)$ must be estimated from observations $x_1, \ldots, x_n \in \mathcal{X}$ and $y_1, \ldots, y_n \in \mathcal{Y}$. For a statistical test we need a statistic and its null distribution. A more explicit expansion of $HSIC(X, Y)$ is as follows.

Proposition 4.6 *$HSIC(X, Y)$ expands as*

$$\begin{aligned}\|m_{XY} - m_X m_Y\|^2 &= \mathbb{E}_{XX'YY'}[k_{\mathcal{X}}(X,X')\, k_{\mathcal{Y}}(Y,Y')] \\ &\quad - 2\,\mathbb{E}_{XY}\big\{\mathbb{E}_{X'}[k_{\mathcal{X}}(X,X')]\; \mathbb{E}_{Y'}[k_{\mathcal{Y}}(Y,Y')]\big\} \\ &\quad + \mathbb{E}_{XX'}[k_{\mathcal{X}}(X,X')]\; \mathbb{E}_{YY'}[k_{\mathcal{Y}}(Y,Y')],\end{aligned}$$

where X and X', and Y and Y', are independent and identically distributed.

[4] For functions of two variables, we use $\cdot$ and $\star$ to indicate the corresponding slots.

Proof See the Appendix at the end of this chapter.

When applying HSIC, we typically replace expectations by empirical averages to obtain the estimator

$$\begin{aligned}\widehat{HSIC} :=& \frac{1}{N^2}\sum_i\sum_j k_{\mathcal{X}}(x_i, x_j)\, k_{\mathcal{Y}}(y_i, y_j)\\ &-\frac{2}{N^3}\sum_i\left\{\sum_j k_{\mathcal{X}}(x_i, x_j)\sum_h k_{\mathcal{Y}}(y_i, y_h)\right\}\\ &+\frac{1}{N^4}\sum_i\sum_j k_{\mathcal{X}}(x_i, x_j)\sum_h\sum_r k_{\mathcal{Y}}(y_h, y_r),\end{aligned} \tag{4.5}$$

where all summations run from 1 to n.

For $x_1, \ldots, x_n \in \mathcal{X}$, define the **centralized Gram matrix** $\tilde{K} = (\tilde{k}_{\mathcal{X}}(x_i, x_j))$ by

$$\tilde{k}_{\mathcal{X}}(x_i, x_j) := \left\langle k_{\mathcal{X}}(x_i, \cdot) - \frac{1}{n}\sum_{s=1}^{n} k_{\mathcal{X}}(x_s, \cdot),\ k_{\mathcal{X}}(x_j, \cdot) - \frac{1}{n}\sum_{t=1}^{n} k_{\mathcal{X}}(x_t, \cdot)\right\rangle.$$

Let E be the $n \times n$ all-ones matrix and I the $n \times n$ identity; set $H := I - \frac{1}{n}E$. Then

$$\tilde{K}_{\mathcal{X}} = HK_{\mathcal{X}}H. \tag{4.6}$$

Indeed, with $\delta_{ij} = 1$ if $i = j$ and 0 otherwise, we have

$$\begin{aligned}&\sum_{s=1}^{n}\sum_{t=1}^{n}(\delta_{is} - \tfrac{1}{n})\, k_{\mathcal{X}}(x_s, x_t)\, (\delta_{tj} - \tfrac{1}{n})\\ =&\sum_{s=1}^{n}(\delta_{is} - \tfrac{1}{n})\left\{k_{\mathcal{X}}(x_s, x_j) - \frac{1}{n}\sum_{t=1}^{n} k_{\mathcal{X}}(x_s, x_t)\right\}\\ =&k_{\mathcal{X}}(x_i, x_j) - \frac{1}{n}\sum_{t=1}^{n} k_{\mathcal{X}}(x_i, x_t) - \frac{1}{n}\sum_{s=1}^{n} k_{\mathcal{X}}(x_s, x_j) + \frac{1}{n^2}\sum_{s=1}^{n}\sum_{t=1}^{n} k_{\mathcal{X}}(x_s, x_t)\\ =&\left\langle k_{\mathcal{X}}(x_i, \cdot) - \frac{1}{n}\sum_{s=1}^{n} k_{\mathcal{X}}(x_s, \cdot),\ k_{\mathcal{X}}(x_j, \cdot) - \frac{1}{n}\sum_{t=1}^{n} k_{\mathcal{X}}(x_t, \cdot)\right\rangle,\end{aligned}$$

where we used $\langle k_{\mathcal{X}}(x_i, \cdot), k_{\mathcal{X}}(x_j, \cdot)\rangle = k_{\mathcal{X}}(x_i, x_j)$. Hence (4.6) holds. ∎

Proposition 4.7 *$\widehat{HSIC}$ can be written as*

$$\widehat{HSIC} = \frac{1}{n^2}\operatorname{tr}(\tilde{K}_{\mathcal{X}}\tilde{K}_{\mathcal{Y}}). \tag{4.7}$$

Proof See the Appendix at the end of this chapter. ■

In Python, the following functions implement these ideas:

```
import numpy as np
import matplotlib.pyplot as plt
from scipy.stats import gaussian_kde
# Function to construct a kernel (Gram) matrix
```

```
def K(k, X):
    X = np.asarray(X)
    if X.ndim == 1:
        n = X.shape[0]
        K_mat = np.zeros((n, n))
        for i in range(n):
            for j in range(n):
                K_mat[i, j] = k(X[i], X[j])
    else:
        n = X.shape[0]
        K_mat = np.zeros((n, n))
        for i in range(n):
            for j in range(n):
                K_mat[i, j] = k(X[i, :], X[j, :])
    return K_mat
```

```
# Function to center a kernel matrix
def tilde(K_mat):
    n = K_mat.shape[0]
    H = np.eye(n) - np.full((n, n), 1.0 / n)
    return H @ K_mat @ H
```

```
# Function to compute HSIC (Hilbert-Schmidt Independence Criterion)
def HSIC_1(K_x, K_y):
    n = K_x.shape[1]
    K_x_tilde = tilde(K_x)
    K_y_tilde = tilde(K_y)
    hsic_value = np.trace(K_x_tilde @ K_y_tilde) / (n ** 2)
    return hsic_value
```

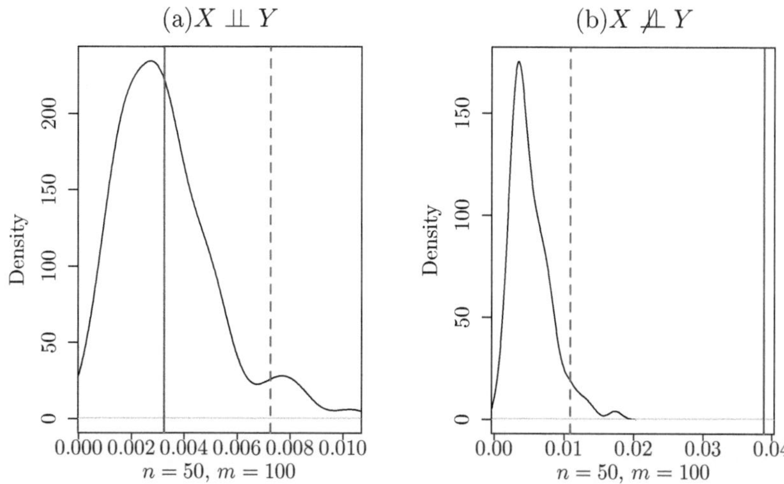

Fig. 4.3 Null distributions (curves) and observed $\widehat{HSIC}$ (blue vertical line) in Example 4.6. The red dashed line marks the rejection boundary. (**a**) Independent X, Y and (**b**) dependent X, Y

For testing independence, the null hypothesis is that X and Y are independent. For the HSIC estimator, a statistic with a known exact null distribution is not available. One may use a **permutation test** or the asymptotic null distribution for large n.

In a permutation test for independence, although $(X, Y) = (x_1, y_1), \ldots, (x_n, y_n)$ may not be independent pairwise, shifted pairs such as $(x_1, y_2), \ldots, (x_{n-1}, y_n)$, (x_n, y_1) are independent. Generate many $\widehat{HSIC}$ values for such independent pairings to form the null distribution. If the observed $\widehat{HSIC}$ lies in the upper α tail, reject the null of independence (one-sided test).

Example 4.6 Using the Gaussian kernel with $\sigma^2 = 1$, we generated pairs of independent standard normal random variables $(x_1, y_1), \ldots, (x_n, y_n)$ $(n = 50$; Fig. 4.3a) and pairs $(x_1, x_1+y_1), \ldots, (x_n, x_n+y_n)$ (Fig. 4.3b). We estimated the null distribution and checked whether the observed $\widehat{HSIC}$ falls in the rejection region. With significance level $\alpha = 0.05$ and $m = 100$ permutations, we did not reject the null for the independent case but did reject for the dependent case. The Python code used is shown below: ∎

```
# Data generation
n = 50
x = np.random.randn(n)
y = np.random.randn(n)

# Define kernel function (RBF kernel)
sigma2 = 1.0
def k_x(a, b): return np.exp(-np.sum((a - b) ** 2) / (2 * sigma2))
k_y = k_x

# Build Gram (kernel) matrices
K_x = K(k_x, x)
K_y = K(k_y, y)

# Compute HSIC value
u = HSIC_1(K_x, K_y)

# Construct the null distribution by permuting x and recomputing
m = 100
w = []
for i in range(m):
    x_shuffled = np.random.permutation(x)
    K_x_perm = K(k_x, x_shuffled)
    w.append(HSIC_1(K_x_perm, K_y))
w = np.array(w)

# Set rejection threshold (significance level 5%)
v = np.quantile(w, 0.95)

# Plot the HSIC null distribution with the observed value
xs = np.linspace(min(w.min(), v, u), max(w.max(), v, u), 400)
kde_w = gaussian_kde(w)
plt.plot(xs, kde_w(xs))
plt.xlim(xs.min(), xs.max())
plt.title("HSIC Null Distribution and Observed Value")
plt.xlabel("HSIC value")
plt.ylabel("Density")
plt.axvline(v, color="red", linestyle="--", linewidth=2)
plt.axvline(u, color="blue", linestyle="-", linewidth=2)
plt.show()
```

Several approaches use asymptotic null distributions. Although Gretton et al. propose a U-statistic approach, here we introduce the method of Kun Zhang [39], which will connect to the next section.

Proposition 4.8 *Let* $\lambda_{\mathcal{X},1} \geq \cdots \geq \lambda_{\mathcal{X},n}$ *and* $\lambda_{\mathcal{Y},1} \geq \cdots \geq \lambda_{\mathcal{Y},n}$ *be the eigenvalues of* $\tilde{K}_{\mathcal{X}}$ *and* $\tilde{K}_{\mathcal{Y}}$*, respectively. Let* Z_{ij}*,* $i, j = 1, \ldots, n$*, be i.i.d.* $N(0, 1)$*. Under* $X \perp\!\!\!\perp Y$*, the statistic* $n\,\widehat{HSIC}$ *is asymptotically distributed as*

$$\frac{1}{n^2}\sum_{i=1}^{n}\sum_{j=1}^{n}\lambda_{\mathcal{X},i}\,\lambda_{\mathcal{Y},j}\,Z_{ij}^2.$$

Proof See the Appendix at the end of this chapter. ∎

Leaving the theory and proofs to the next section, we now conduct numerical experiments using this method. We extend the function `HSIC_1` to also compute eigenvalues and return n^3 times the statistic (i.e., n times the trace) together with the eigenvalues.

```
# Function that returns n-times-cubed HSIC value and eigenvalues (following
   the original implementation)
def HSIC_2(K_x, K_y):
    n = K_x.shape[1]
    tilde_x = tilde(K_x)
    tilde_y = tilde(K_y)
    hsic_value = np.trace(tilde_x @ tilde_y)
    lambda_x = np.real(np.linalg.eigvals(tilde_x))
    lambda_y = np.real(np.linalg.eigvals(tilde_y))
    lambda_xy = np.sort((lambda_x[:, None] * lambda_y [None, :]).ravel())
      [::-1]
    return {"statistics": hsic_value * n, "eigen_values": lambda_xy}
```

Example 4.7 We computed the eigenvalues of $\tilde{K}_{\mathcal{X}}$ and $\tilde{K}_{\mathcal{Y}}$ and, following Proposition 4.8, generated 10,000 random values from the distribution of $n\,\widehat{HSIC}$ to obtain the upper $\alpha = 0.05$ critical value. As in Example 4.6, we generated data for the independent and dependent cases. The null was not rejected in the independent case and was rejected in the dependent case. The function `null_dist` below returns the simulated values z of $\sum_{i=1}^{n}\sum_{j=1}^{n}\lambda_{\mathcal{X},i}\lambda_{\mathcal{Y},j}Z_{ij}^2$ and its $100(1-\alpha)$th percentile. ■

```
# Function that sets the rejection region
def null_dist(eigen, alpha):
    if len(eigen) == 0:
        return {"z": np.array([]), "critical": np.nan}
    r = len(eigen)
    z = np.empty(10000)
    for h in range(10000):
        z[h] = np.sum(eigen * np.random.chisquare(df=1, size=r))
    critical = np.quantile(z, 1 - alpha)
    return {"z": z, "critical": critical}
```

Verification code (note that `null_dist` outputs the unscaled $\sum \lambda_{\mathcal{X},i}\lambda_{\mathcal{Y},j}Z_{ij}^2$):

```
# Set sample size and generate data
n = 50
x = np.random.randn(n)
y = np.random.randn(n)
# y = x + np.random.randn(n)

# Compute kernel Gram matrices
K_x = K(k_x, x)
K_y = K(k_y, y)

# Compute HSIC and obtain eigenvalues
result = HSIC_2(K_x, K_y)
actual_value = result["statistics"]
eigen = result["eigen_values"]

# Compute the 95th percentile of the HSIC null distribution
critical = null_dist(eigen, 0.05)["critical"]

# Display results
print("95th percentile of HSIC null distribution:", critical)
print("n times the observed HSIC value:", actual_value)
```

Example output for independent standard normals X, Y:

```
> 95% critical point of the null: 621.8676
> Observed n * HSIC: 253.9742
```

When setting `y = x + rng.standard_normal(n)`:

```
> 95% critical point of the null: 743.1694
> Observed n * HSIC: 4964.367
```

4.3 Testing Conditional Independence: KCI

Let $\mathcal{X}$, $\mathcal{Y}$, and $\mathcal{Z}$ be sets, and suppose we have observations $(X, Y, Z) = (x_1, y_1, z_1), \ldots, (x_n, y_n, z_n) \in \mathcal{X} \times \mathcal{Y} \times \mathcal{Z}$. We test conditional independence $X \perp\!\!\!\perp Y \mid Z$. In this section, we introduce the **Kernel-based Conditional Independence** (KCI) test proposed by Kun Zhang et al. in 2011 [39].

Let $K_{\mathcal{Z}} \in \mathbb{R}^{n \times n}$ be the Gram matrix of a kernel $k_{\mathcal{Z}} : \mathcal{Z} \times \mathcal{Z} \to \mathbb{R}$. Let $k_{XZ} : \mathcal{X} \times \mathcal{Z} \times \mathcal{X} \times \mathcal{Z} \to \mathbb{R}$ be the product kernel of k_X and k_Z, and let $K_{\mathcal{X}\mathcal{Z}} \in \mathbb{R}^{n \times n}$ be its Gram matrix. Define the centered matrices $\tilde{K}_{\mathcal{Z}} := HK_{\mathcal{Z}}H$ and $\tilde{K}_{\mathcal{X}\mathcal{Z}} := HK_{\mathcal{X}\mathcal{Z}}H$. Define $\tilde{K}_{\mathcal{Y}\mathcal{Z}}$ analogously.

To construct a test statistic T for conditional independence, we introduce **kernel ridge regression**. Given data $x_1, \ldots, x_n \in \mathcal{X} = \mathbb{R}$ and $y_1, \ldots, y_n \in \mathbb{R}$, for $\lambda > 0$ consider

$$\sum_{i=1}^{n} \{y_i - f(x_i)\}^2 + \lambda \|f\|_H^2, \tag{4.8}$$

where H is an RKHS with norm $\|\cdot\|_H$ and kernel k. If $f \in H$, the solution is of the form $f(x) = \sum_{j=1}^{n} \alpha_j k(x, x_j)$ for some $\alpha = [\alpha_1, \ldots, \alpha_n]^\top$, with $\hat{\alpha} = (K + \lambda I)^{-1}[y_1, \ldots, y_n]^\top$ (Problem 38).

In matrix form,

$$\begin{bmatrix} \hat{f}(x_1) \\ \vdots \\ \hat{f}(x_n) \end{bmatrix} = \begin{bmatrix} \sum_{j=1}^{n} \hat{\alpha}_j k(x_1, x_j) \\ \vdots \\ \sum_{j=1}^{n} \hat{\alpha}_j k(x_n, x_j) \end{bmatrix} = K\hat{\alpha} = K(K + \lambda I)^{-1} \begin{bmatrix} y_1 \\ \vdots \\ y_n \end{bmatrix}.$$

The penalty term proportional to $\|f\|_H$ in (4.8) prevents overfitting by discouraging overly complex functions. With $\lambda = 0$ (ordinary least squares), the prediction $f(X)$ may be overly sensitive to X. Choosing $\lambda > 0$ appropriately yields predictions closer to the average response given X. Note how the $(K + \lambda I)^{-1}$ factor shrinks coefficients compared to K^{-1}.

We now apply this idea not to $\mathcal{X}$ and $\mathcal{Y}$ directly, but to

$$(\mathcal{X} \times \mathcal{Z}) \ni (x, z) \mapsto k_{\mathcal{X}\mathcal{Z}}((x, z), (\cdot, \star)) \in H_{\mathcal{X}\mathcal{Z}},$$

$$(\mathcal{Y} \times \mathcal{Z}) \ni (y, z) \mapsto k_{\mathcal{Y}\mathcal{Z}}((y, z), (\cdot, \star)) \in H_{\mathcal{Y}\mathcal{Z}},$$

and test conditional independence in $H_{\mathcal{XZ}}$ and $H_{\mathcal{YZ}}$.[5] KCI uses the centered kernels $\tilde{k}_{\mathcal{XZ}}$ and $\tilde{k}_{\mathcal{YZ}}$. For $f \in H_{\mathcal{XZ}}$ and $g \in H_{\mathcal{YZ}}$, subtract the conditional mean given $Z = z$:

$$f(x,z) - \tilde{K}_{\mathcal{Z}}(\tilde{K}_{\mathcal{Z}} + \lambda I)^{-1} f(x,z) = R_{\mathcal{Z}}\, f(x,z),$$

$$g(y,z) - \tilde{K}_{\mathcal{Z}}(\tilde{K}_{\mathcal{Z}} + \lambda I)^{-1} g(y,z) = R_{\mathcal{Z}}\, g(y,z),$$

where

$$R_{\mathcal{Z}} := I - \tilde{K}_{\mathcal{Z}}(\tilde{K}_{\mathcal{Z}} + \lambda I)^{-1} = \lambda(\tilde{K}_{\mathcal{Z}} + \lambda I)^{-1}.$$

Thus the feature maps

$$(x,z) \mapsto \begin{bmatrix} \tilde{k}_{\mathcal{XZ}}((x_1,z_1),(x,z)) \\ \vdots \\ \tilde{k}_{\mathcal{XZ}}((x_n,z_n),(x,z)) \end{bmatrix}, \qquad (y,z) \mapsto \begin{bmatrix} \tilde{k}_{\mathcal{YZ}}((y_1,z_1),(y,z)) \\ \vdots \\ \tilde{k}_{\mathcal{YZ}}((y_n,z_n),(y,z)) \end{bmatrix}$$

are replaced by

$$(x,z) \mapsto R_{\mathcal{Z}} \begin{bmatrix} \tilde{k}_{\mathcal{XZ}}((x_1,z_1),(x,z)) \\ \vdots \\ \tilde{k}_{\mathcal{XZ}}((x_n,z_n),(x,z)) \end{bmatrix}, \qquad (y,z) \mapsto R_{\mathcal{Z}} \begin{bmatrix} \tilde{k}_{\mathcal{YZ}}((y_1,z_1),(y,z)) \\ \vdots \\ \tilde{k}_{\mathcal{YZ}}((y_n,z_n),(y,z)) \end{bmatrix}.$$

Hence, with $\tilde{K}_{\mathcal{XZ}} = V_{\mathcal{XZ}} V_{\mathcal{XZ}}^\top$ and $\tilde{K}_{\mathcal{YZ}} = V_{\mathcal{YZ}} V_{\mathcal{YZ}}^\top$,

$$\tilde{K}_{\mathcal{XZ}|\mathcal{Z}} = R_{\mathcal{Z}} V_{\mathcal{XZ}} \{R_{\mathcal{Z}} V_{\mathcal{XZ}}\}^\top = R_{\mathcal{Z}}\, \tilde{K}_{\mathcal{XZ}}\, R_{\mathcal{Z}},$$

$$\tilde{K}_{\mathcal{YZ}|\mathcal{Z}} = R_{\mathcal{Z}} V_{\mathcal{YZ}} \{R_{\mathcal{Z}} V_{\mathcal{YZ}}\}^\top = R_{\mathcal{Z}}\, \tilde{K}_{\mathcal{YZ}}\, R_{\mathcal{Z}}.$$

Analogously to the independence test, define the statistic

$$T := \frac{1}{n} \operatorname{tr}\left(\tilde{K}_{\mathcal{XZ}|\mathcal{Z}}\, \tilde{K}_{\mathcal{YZ}|\mathcal{Z}}\right),$$

and derive its asymptotic null distribution under $X \perp\!\!\!\perp Y \mid Z$.[6]

---

[5] If $k : \mathcal{X} \times \mathcal{X} \to \mathbb{R}$ and $x \in \mathcal{X}$, then $k(x,\cdot)$ denotes the function $y \mapsto k(x,y)$. For $k : (\mathcal{X} \times \mathcal{Y}) \times (\mathcal{X} \times \mathcal{Y}) \to \mathbb{R}$ and $(x,y) \in \mathcal{X} \times \mathcal{Y}$, we write $k(x,y,\cdot,\star)$ for the function $(u,v) \mapsto k(x,y,u,v)$.

[6] The original KCI paper uses $T := \frac{1}{n}\operatorname{tr}(\tilde{K}_{\mathcal{XZ}|\mathcal{Z}} \tilde{K}_{\mathcal{Y}|\mathcal{Z}})$ and derives the null with $\mathcal{YZ}$ replaced by $\mathcal{Y}$. This is equivalent to the treatment here. The original form is computationally convenient; we adopt a symmetric form for clarity.

Let $R_\mathcal{Z} V_{\mathcal{X}\mathcal{Z}} = \phi = (\phi_{ij})$ and $R_\mathcal{Z} V_{\mathcal{Y}\mathcal{Z}} = \varphi = (\varphi_{ik})$, so that $\tilde{K}_{\mathcal{X}\mathcal{Z}|\mathcal{Z}} = \phi\phi^\top$ and $\tilde{K}_{\mathcal{Y}\mathcal{Z}|\mathcal{Z}} = \varphi\varphi^\top$. Define $w_i = (w_{i,j,k}) \in \mathbb{R}^{n^2}$ with entries $w_{i,j,k} = \phi_{ij}\varphi_{ik}$, and form the matrix $w = [w_1, \ldots, w_n] \in \mathbb{R}^{n^2 \times n}$. Then each entry of $ww^\top \in \mathbb{R}^{n^2 \times n^2}$ at index $((j,k),(j',k'))$ is

$$\sum_{i=1}^{n} \phi_{ij}\varphi_{ik}\phi_{ij'}\varphi_{ik'}.$$

Let $\lambda_1, \ldots, \lambda_{n^2}$ be the eigenvalues of $ww^\top$.

Proposition 4.9 (Kun Zhang 2011) *Let $Z_1, \ldots, Z_{n^2} \sim N(0,1)$ be independent. Under $X \perp\!\!\!\perp Y \mid Z$ and as $n \to \infty$, the statistic T is distributed as*

$$\frac{1}{n}\sum_{k=1}^{n^2} \lambda_k Z_k^2.$$

Proof See the Appendix at the end of this chapter. ■

A Python implementation is as follows. We first force symmetry to avoid numerical issues (e.g., Cholesky failures) when matrices are not exactly symmetric.

```
def symmetrize(A):
    return (A.T + A) / 2.0
```

```
# KCI (Kernel-based Conditional Independence Test)
def KCI_1(K_x, K_y, K_z, lambda_):
    n = K_x.shape[0]
    K_xz = K_x * K_z
    K_yz = K_y * K_z
    R_Z = lambda_ * np.linalg.inv(tilde(K_z) + lambda_ * np.eye(n))
    XZ = R_Z @ tilde(K_xz) @ R_Z.T
    YZ = R_Z @ tilde(K_yz) @ R_Z.T
    eigen_XZ = np.linalg.eig(symmetrize(XZ))
    eigen_YZ = np.linalg.eig(symmetrize(YZ))
    r = int(np.ceil(np.sqrt(n)))
    S = eigen_XZ[1][:, :r] @ np.diag(np.sqrt(np.real(eigen_XZ[0][:r])))
    T = eigen_YZ[1][:, :r] @ np.diag(np.sqrt(np.real(eigen_YZ[0][:r])))
    W = np.zeros((n, r * r))
    for i in range(r):
        for j in range(r):
            W[:, r * i + j] = S[:, i] * T[:, j]
    return {
        "statistics": np.trace(XZ @ YZ),
        "eigen_values": np.real(np.linalg.eigvals(W.T @ W)),
    }
```

Up to lines 7 and 8 we have

$$\tilde{K}_{\mathcal{X}\mathcal{Z}|\mathcal{Z}} = \phi\phi^\top, \qquad \tilde{K}_{\mathcal{Y}\mathcal{Z}|\mathcal{Z}} = \varphi\varphi^\top.$$

In lines 9 and 10 we write

$$\phi = U\Lambda_X^{1/2}, \quad U = [u_1, \ldots, u_n] = \begin{bmatrix} u_{11} & \cdots & u_{1n} \\ \vdots & \ddots & \vdots \\ u_{n1} & \cdots & u_{nn} \end{bmatrix}, \quad \Lambda_X = \mathrm{diag}(\lambda_{X,1}, \ldots, \lambda_{X,n}),$$

$$\varphi = V\Lambda_Y^{1/2}, \quad V = [v_1, \ldots, v_n] = \begin{bmatrix} v_{11} & \cdots & v_{1n} \\ \vdots & \ddots & \vdots \\ v_{n1} & \cdots & v_{nn} \end{bmatrix}, \quad \Lambda_Y = \mathrm{diag}(\lambda_{Y,1}, \ldots, \lambda_{Y,n}),$$

so that

$$\phi_{ij} = \sqrt{\lambda_{X,j}}\, u_{ij}, \qquad \varphi_{ik} = \sqrt{\lambda_{Y,k}}\, v_{ik}, \qquad i, j, k = 1, \ldots, n.$$

Thus arrays `S` and `T` (lines 12 and 13) store ϕ_{ij} and φ_{ik}, respectively; in `W` (lines 14-17), the two-dimensional index (j, k) is flattened to one dimension so that $w_{i,j,k} = \phi_{ij}\varphi_{ik}$ is stored as `W[i, r*(i-1)+j]`.

The eigenvalues decay roughly exponentially, and computing eigenvalues of an $n^2 \times n^2$ matrix is often infeasible. Therefore, we select the top r eigenvalues $\lambda_{X,1}, \ldots, \lambda_{X,n}$ and $\lambda_{Y,1}, \ldots, \lambda_{Y,n}$ and the corresponding eigenvectors $u_1, \ldots, u_r$ and $v_1, \ldots, v_r$, reducing W to size $n \times r^2$. The results are nearly unchanged compared to $r = n$.

Example 4.8 Let U, V, W be independent random variables taking values ± 1 with equal probability, and define X, Y, Z as follows. Generate $n = 100$ samples and test $X \perp\!\!\!\perp Y \mid Z$.

1. $X = U, Y = V, Z = W$.
2. $X = UW, Y = VW, Z = W$.
3. $X = U, Y = V, Z = UV$.
4. $X = U, Y = UV, Z = V$.

With $\lambda = 0.1$ and $r = 10$, we ran the program below. The statistic and null distributions for each case are shown in Fig. 4.4. In (a) and (b) the statement $X \perp\!\!\!\perp Y \mid Z$ is true, whereas in (c) and (d) it is false. The statistical test accepts the null in (a) and (b) and rejects it in (c) and (d). ■

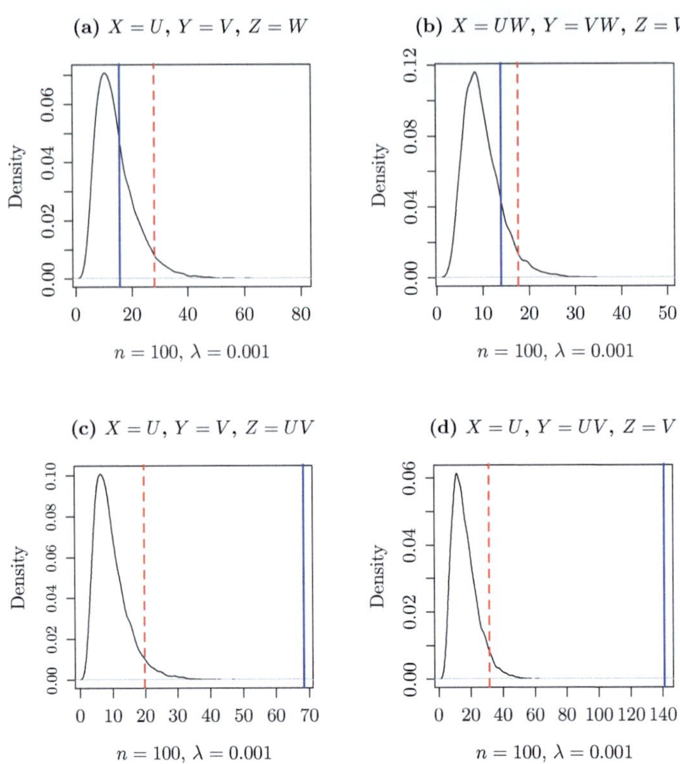

Fig. 4.4 Conditional Independence Tests for Example 4.8

```
# Kernel setup (Gaussian/RBF kernel)
def k_x(a, b): return np.exp(-np.sum((a - b) ** 2) / 2)
k_y = k_x
k_z = k_x

# Generate data
n = 100
x = np.random.randn(n)
y = np.random.randn(n)
z = np.random.randn(n)

# Compute kernel Gram matrices
K_x = K(k_x, x)
K_y = K(k_y, y)
K_z = K(k_z, z)

# Set parameters and run KCI
lambda_ = 0.001
result = KCI_1(K_x, K_y, K_z, lambda_)
eigen = result["eigen_values"]
u = result["statistics"]

alpha = 0.05
result2 = null_dist(eigen, alpha)
z_vals = result2["z"]
v = result2["critical"]
```

```
# Plot results
xs = np.linspace(min(z_vals.min(), v, u), max(z_vals.max(), v, u), 400)
kde_z = gaussian_kde(z_vals)
plt.plot(xs, kde_z(xs))
plt.xlim(xs.min(), xs.max())
plt.title("KCI null distribution and observed statistic")
plt.xlabel("Statistic")
plt.ylabel("Density")
plt.axvline(v, color="red", linestyle="--", linewidth=2)
plt.axvline(u, color="blue", linestyle="-", linewidth=2)
plt.show()
```

If the regularization parameter λ is too small, computing $(\tilde{K}_{\mathcal{Z}} + \lambda I)^{-1}$ can become unstable, and numerical issues—such as slightly negative eigenvalues due to rounding error—may occur. In this book's Python implementation we use the defaults $\lambda = 10^{-3}$ and significance level $\alpha = 0.05$ (the same settings as the R package `CondIndTests`). For numerical stabilization, we symmetrize via $A \mapsto (A + A^\top)/2$ and clip eigenvalues at zero with $\max(\cdot, 0)$. When needed, adjust λ within $10^{-2} \sim 10^{-4}$, and add a small diagonal jitter ϵI (with $\epsilon \approx 10^{-8}$) for extra stability.

Note that the RBF kernel bandwidth (σ^2) should ideally be selected by cross-validation; as a simple alternative, the median heuristic can be used. Also, while we use the Gaussian kernel with $\sigma^2 = 1$ here, in principle σ^2 should be tuned (see Problem 36). In Chap. 5, we adopt the same parameter settings as `CondIndTests`.

Appendix

Proof of Proposition 4.6

Since

$$HSIC(X, Y) = \|m_{XY}\|^2 - 2\langle m_{XY}, m_X m_Y\rangle + \|m_X m_Y\|^2,$$

we compute each term as follows:

$$\begin{aligned}
\|m_{XY}\|^2 &= \big\langle \mathbb{E}_{XY}[k_{\mathcal{X}}(X,\cdot)k_{\mathcal{Y}}(Y,\cdot)],\ \mathbb{E}_{X'Y'}[k_{\mathcal{X}}(X',\cdot)k_{\mathcal{Y}}(Y',\cdot)]\big\rangle \\
&= \mathbb{E}_{XY}\mathbb{E}_{X'Y'}\big[\langle k_{\mathcal{X}}(X,\cdot)k_{\mathcal{Y}}(Y,\cdot),\ k_{\mathcal{X}}(X',\cdot)k_{\mathcal{Y}}(Y',\cdot)\rangle\big] \\
&= \mathbb{E}_{XX'YY'}[k_{\mathcal{X}}(X,X')\,k_{\mathcal{Y}}(Y,Y')], \\
\langle m_{XY}, m_X m_Y\rangle &= \big\langle \mathbb{E}_{XY}[k_{\mathcal{X}}(X,\cdot)k_{\mathcal{Y}}(Y,\cdot)],\ \mathbb{E}_{X'}[k_{\mathcal{X}}(X',\cdot)]\,\mathbb{E}_{Y'}[k_{\mathcal{Y}}(Y',\cdot)]\big\rangle \\
&= \mathbb{E}_{XY}\big\{\mathbb{E}_{X'}\big[\langle k_{\mathcal{X}}(X,\cdot)k_{\mathcal{Y}}(Y,\cdot),\ k_{\mathcal{X}}(X',\cdot)\,\mathbb{E}_{Y'}[k_{\mathcal{Y}}(Y',\cdot)]\rangle\big]\big\} \\
&= \mathbb{E}_{XY}\big\{\mathbb{E}_{X'}[k_{\mathcal{X}}(X,X')]\,\mathbb{E}_{Y'}[k_{\mathcal{Y}}(Y,Y')]\big\},
\end{aligned}$$

$$\|m_X m_Y\|^2 = \langle \mathbb{E}_X[k_\mathcal{X}(X,\cdot)]\,\mathbb{E}_Y[k_\mathcal{Y}(Y,\cdot)],\ \mathbb{E}_{X'}[k_\mathcal{X}(X',\cdot)]\,\mathbb{E}_{Y'}[k_\mathcal{Y}(Y',\cdot)]\rangle$$
$$= \mathbb{E}_X\mathbb{E}_{X'}[k_\mathcal{X}(X,X')]\,\mathbb{E}_Y\mathbb{E}_{Y'}[k_\mathcal{Y}(Y,Y')].$$

■

Proof of Proposition 4.7

Using $\mathrm{tr}(HK_\mathcal{X}H \cdot HK_\mathcal{Y}H) = \mathrm{tr}(K_\mathcal{X}H \cdot HK_\mathcal{Y}H \cdot H)$ and $H^2 = H$, we have

$$\mathrm{tr}(\tilde{K}_\mathcal{X}\tilde{K}_\mathcal{Y}) = \mathrm{tr}(K_\mathcal{X}HK_\mathcal{Y}H),$$

which equals $\widehat{HSIC}$ as follows:

$$\mathrm{tr}(K_\mathcal{X}HK_\mathcal{Y}H) = \sum_i (K_\mathcal{X}HK_\mathcal{Y}H)_{ii} = \sum_i\sum_j (K_\mathcal{X}H)_{ij}\,(K_\mathcal{Y}H)_{ji}$$
$$= \sum_i\sum_j \left\{\sum_h k_\mathcal{X}(x_i,x_h)(\delta_{hj}-\tfrac{1}{n})\right\}\left\{\sum_h k_\mathcal{Y}(y_j,y_h)(\delta_{hi}-\tfrac{1}{n})\right\}$$
$$= \sum_i\sum_j \left\{k_\mathcal{X}(x_i,x_j)k_\mathcal{Y}(y_i,y_j) - \frac{1}{n}k_\mathcal{X}(x_i,x_j)\sum_h k_\mathcal{Y}(y_i,y_h)\right.$$
$$\left. - \frac{1}{n}k_\mathcal{Y}(y_i,y_j)\sum_h k_\mathcal{X}(x_i,x_h) + \frac{1}{n^2}\sum_h k_\mathcal{X}(x_i,x_h)\sum_r k_\mathcal{Y}(y_j,y_r)\right\}$$
$$= \sum_i\sum_j k_\mathcal{X}(x_i,x_j)k_\mathcal{Y}(y_i,y_j) - \frac{2}{n}\sum_i\sum_j k_\mathcal{X}(x_i,x_j)\sum_h k_\mathcal{Y}(y_i,y_h)$$
$$+ \frac{1}{n^2}\sum_i\sum_h k_\mathcal{X}(x_i,x_h)\sum_j\sum_r k_\mathcal{Y}(y_j,y_r).$$

■

Proofs of Propositions 4.8 and 4.9

Let $\phi = (\phi_{ij})$ and $\varphi = (\varphi_{ik})$, and set $w_{i,j,k} := \phi_{ij}\varphi_{ik}$. Then

$$\mathrm{tr}\Big(\tilde{K}_{\mathcal{X}\mathcal{Z}|\mathcal{Z}}\,\tilde{K}_{\mathcal{Y}\mathcal{Z}|\mathcal{Z}}\Big) = \mathrm{tr}(\phi\phi^\top\varphi\varphi^\top) = \mathrm{tr}(\phi^\top\varphi\,\varphi^\top\phi) = \mathrm{tr}\big((\phi^\top\varphi)(\phi^\top\varphi)^\top\big)$$
$$= \|\phi^\top\varphi\|_F^2 = \sum_{j=1}^n\sum_{k=1}^n\left(\sum_{i=1}^n \phi_{ij}\varphi_{ik}\right)^2 = \sum_{j=1}^n\sum_{k=1}^n\left(\sum_{i=1}^n w_{i,j,k}\right)^2,$$

where $\|A\|_F$ denotes the Frobenius norm $\sqrt{\sum_i \sum_j a_{ij}^2}$. We used $\mathrm{tr}(AB) = \mathrm{tr}(BA)$ when both products are defined and $\mathrm{tr}(AA^\top) = \|A\|_F^2$ (Problem 41). Hence set

$$S_n := \frac{1}{\sqrt{n}} \sum_{i=1}^{n} w_i \in \mathbb{R}^{n^2}, \qquad T = \|S_n\|^2.$$

We use the following lemma.

Lemma 4.1 (Daudin 1980 [6]) *Let $f : \mathcal{X} \times \mathcal{Z} \to \mathbb{R}$ and $g : \mathcal{Y} \times \mathcal{Z} \to \mathbb{R}$ be square-integrable with $\mathbb{E}_X[f \mid Z] = 0$ and $\mathbb{E}_Y[g \mid Z] = 0$. Then, almost surely in Z,*

$$\mathbb{E}_{XY}[f(X, Z)g(Y, Z)] = 0$$

if and only if $X \perp\!\!\!\perp Y \mid Z$.[7]

Since $\sum_{i=1}^n \phi_{ij} = \sum_{i=1}^n \varphi_{ik} = 0$ (Problem 36), under the null $X \perp\!\!\!\perp Y \mid Z$ the vector $w_i = (\phi_{ij}\varphi_{ik})$ has mean zero by Lemma 4.1. By the multivariate central limit theorem, S_n is asymptotically normal with mean zero, and by the law of large numbers, $\frac{1}{n} ww^\top$ converges to the covariance matrix $\Sigma \in \mathbb{R}^{n^2 \times n^2}$. Since T is the sum of squares of the components of S_n, left-multiplying by an orthogonal matrix $P \in \mathbb{R}^{n^2 \times n^2}$ does not change T. We can diagonalize Σ to $\mathrm{diag}(\lambda_1/n, \ldots, \lambda_{n^2}/n)$. Therefore, under the null and for large n,

$$T \sim \frac{1}{n} \sum_{k=1}^{n^2} \lambda_k Z_k^2,$$

with independent $Z_k \sim N(0, 1)$.

Finally, we show that Proposition 4.8 is a special case of Proposition 4.9. The eigenvalues of the null distribution are products $\lambda_{X,j}\lambda_{Y,k}$. With $\phi_{ij} = \sqrt{\lambda_{X,j}} u_{ij}$ and $\varphi_{ik} = \sqrt{\lambda_{Y,k}} v_{ik}$, the $((j, k), (j', k'))$ entry of Σ is

$$\frac{1}{n} \sum_{i=1}^{n} \sqrt{\lambda_{X,j}\lambda_{X,j'}} \sqrt{\lambda_{Y,k}\lambda_{Y,k'}}\, u_{ij}u_{ij'}\, v_{ik}v_{ik'}.$$

Under the null and asymptotically,

$$\frac{1}{n} \sum_{i=1}^{n} \sqrt{\lambda_{X,j}\lambda_{X,j'}}\, u_{ij}u_{ij'} \cdot \frac{1}{n} \sum_{i'=1}^{n} \sqrt{\lambda_{Y,k}\lambda_{Y,k'}}\, v_{i'k}v_{i'k'}$$

so that diagonal entries converge to $\lambda_{X,j}\lambda_{Y,k}$ and off-diagonals to 0, using $\sum_{i=1}^n u_{ij}u_{ij'} = \delta_{jj'}$ and $\sum_{i=1}^n v_{ik}v_{ik'} = \delta_{kk'}$. ■

[7] It suffices if one of f or g depends only on X or Y with zero mean.

Problems 30–43

30. I would like to implement the Nadaraya–Watson estimator and obtain an output like Fig. 4.1. Construct (i) the Epanechnikov kernel function D, (ii) a function K that returns the value of the Gaussian kernel for $x, y \in \mathbb{R}^n$ and $\sigma^2 > 0$, and (iii) a function f that predicts y_* at a new value $x_* \in \mathbb{R}$ with $\sigma^2 > 0$ via the Nadaraya–Watson estimator. Apply them to the program below and verify its behavior.

```
def D(t):
    """
␣␣␣␣Epanechnikov␣kernel:␣D(t)␣=␣3/4␣*␣(1␣-␣t^2)␣*␣1(|t|␣<=␣1)
␣␣␣␣t:␣scalar␣or␣array
␣␣␣␣"""

def K(x, y, sigma2):
    """
␣␣␣␣Gaussian␣(RBF)␣kernel:␣K(x,␣y)␣=␣exp(␣-||x␣-␣y||^2␣/␣(2␣*␣sigma^2)␣)
␣␣␣␣x,␣y:␣vectors␣in␣R^n␣(ndarray)
␣␣␣␣sigma2:␣positive␣real␣number␣(variance)
␣␣␣␣"""

def f(x_star, lam, x, y):
    """
␣␣␣␣Nadaraya--Watson␣estimator␣(Epanechnikov␣kernel,␣bandwidth␣lam)
␣␣␣␣x_star:␣target␣x␣value␣to␣predict␣at␣(scalar)
␣␣␣␣lam:␣bandwidth␣?␣>␣0
␣␣␣␣x,␣y:␣observed␣data␣(1D␣arrays␣of␣the␣same␣length)
␣␣␣␣"""

# --- Data generation (same idea as the R code) ---
n = 250
rng = np.random.default_rng(0)
x = 2.0 * rng.standard_normal(n)
y = np.sin(2.0 * np.pi * x) + rng.standard_normal(n) / 4.0

# --- Plotting ---
plt.figure()
plt.plot([], [])  # An axis-only initialization equivalent to R's type='
   n' (for axis labels)
plt.xlim(-3, 3)
plt.ylim(-2, 3)
plt.xlabel("x")
plt.ylabel("y")

# Scatter plot
plt.scatter(x, y, s=10, c="black", alpha=0.6)

# Estimated curves
xx = np.arange(-3, 3.0 + 1e-12, 0.05)
lambdas = [0.05, 0.35, 0.50]
colors = ["green", "blue", "red"]

for lam, c in zip(lambdas, colors):
    yy = np.array([f(z, lam, x, y) for z in xx])
    plt.plot(xx, yy, color=c, linewidth=2, label=f"?={lam}")

plt.legend()
plt.tight_layout()
plt.show()
```

31. Show that at least one eigenvalue of the Gram matrix K_λ in Example 4.1 is negative, and hence K_λ is not positive semidefinite. *Hint:* For a symmetric matrix $A \in \mathbb{R}^{N\times N}$, A is positive semidefinite if and only if all its eigenvalues are nonnegative.
32. Show that the Gaussian kernel and the Cauchy kernel are, respectively, characteristic functions of the corresponding distributions. Also, for $\mathcal{X} = \mathbb{R}$, write Python functions that output the values of the Gaussian kernel with $\sigma^2 = 1$ and the Cauchy kernel with $\beta = 1$.
33. Show that the inner product of a finite-dimensional Euclidean space defines a kernel.
34. Let k be a positive definite kernel. Show that the set $H_0 = \{f = \sum_{i=1}^{N} \alpha_i k(x_i, \cdot) : N \geq 1,\ x_1, \ldots, x_N \in \mathcal{X},\ \alpha_1, \ldots, \alpha_N \in \mathbb{R}\}$ satisfies (4.1). Also show that the inner product $\sum_{i=1}^{N}\sum_{j=1}^{N} \alpha_i \beta_j\, k(x_i, x_j)$ for $f = \sum_{i=1}^{N} \alpha_i k(x_i, \cdot)$ and $g = \sum_{j=1}^{N} \beta_j k(x_j, \cdot) \in H_0$ satisfies (4.2).
35. Show that L^2 in Example 4.4 is a linear space. Also, state how the norm $\|f\|$ of each element f is defined.
36. Using the fact that each row sum and each column sum of the centered Gram matrix $\tilde{K}$ equals 0, show that $\displaystyle\sum_{i=1}^{n} \phi_{i,j} = \sum_{i=1}^{n} \varphi_{i,j} = 0$. *Hint:* Derive $1^\top R_{\mathcal{Z}} V_{\mathcal{X}\mathcal{Z}} = 1^\top V_{\mathcal{X}\mathcal{Z}} = 0$.
37. For the datasets below, display the relationship between the null distribution and the test statistic in the same format as the figure in Example 4.7.

```
x = rng.standard_normal(n); y = rng.standard_normal(n)
x = rng.standard_normal(n); y = x + rng.standard_normal(n)
```

38. Write $f \in H$ as the sum of a linear combination of $k(x_i, \cdot)$, $i = 1, \ldots, n$, namely $\sum_{i=1}^{n} \alpha_i k(x_i, \cdot)$, and a function $f_\perp$ orthogonal to that span. Then:

    (a) Show that $\|f\|^2 = \left\|\sum_{i=1}^{n} \alpha_i k(x_i, \cdot)\right\|^2 + \|f_\perp\|^2$.
    (b) Show that the minimizer $f \in H$ of (4.8) can be expressed as a linear combination of $k(x_i, \cdot)$ for $i = 1, \ldots, n$.

39. Using the same data as in Example 4.8 (all four cases), change the parameters to $\lambda = 0.01$ and $r = 3, 20$, run the procedure, and check whether the results change.
40. In Example 4.8, let each random variable follow the standard normal distribution, and define:

    (a) $X = U, Y = V, Z = W$.
    (b) $X = U + W, Y = V + W, Z = W$.
    (c) $X = U, Y = V, Z = U + V$.
    (d) $X = U, Y = U + V, Z = V$.

    For each case, display the test statistic and the null distribution in the style of Fig. 4.4.

41. Show that $\text{tr}(AB) = \text{tr}(BA)$ whenever both products AB and BA are defined and that for a matrix $A = (a_{ij})$, $\text{tr}(AA^\top) = \|A\|_F^2$.
42. Unconditional independence corresponds to Z taking a fixed value. Suppose $k_{\mathcal{Z}}(z, z')$ is a constant $a > 0$ for all $z, z' \in \mathcal{Z}$. Show the following:

    (a) $R_{\mathcal{Z}}$ is the identity matrix.

    (b) Both T and $\frac{1}{n}\sum_{k=1}^{n^2} \lambda_k Z_k^2$ are multiplied by a^2.

43. In Example 2.9 we presented a case where the correlation between X and Y is zero, but they are not independent. In such cases, nonindependence cannot be detected by the correlation coefficient. Generate $n = 100$ random samples of the form (x_i, x_i^2), $i = 1, \ldots, n$ and test independence using HSIC (significance level $\alpha = 0.01$). Furthermore, let Z be a binary random variable taking values 0 and 1 with equal probability. When $Z = 0$, set $Y = X^2$; when $Z = 1$, set $X = Y^2$. Generate $n = 100$ random triples (x_i, y_i, z_i) and test conditional independence using KCI.

Chapter 5
PC Algorithm

In this chapter, we explain the PC algorithm, which constructs a DAG based on tests of conditional independence. The algorithm enables efficient structure estimation, especially for high-dimensional data.

We begin with the faithfulness assumption that underpins the theoretical validity of the PC algorithm. This means that all conditional independences in a probability distribution can be represented by d-separation in some DAG. Next, assuming the Python implementations prepared in Chap. 4 (HSIC/KCI, etc.), we outline the overall flow of the PC algorithm and its implementation details.

We then present the first half of the algorithm, which decides the presence or absence of edges using general conditional independence (CI) tests such as KCI. Along with a Python example, we concretely describe how the undirected skeleton is constructed.

Finally, based on rules that detect colliders and exclude directed cycles, we explain the second half, which orients edges and produces a CPDAG. We also discuss the situation where complete orientation is impossible and a CPDAG (allowing some undirected edges) is obtained.

At the end of the chapter, we apply what we have developed to real data.

(Programs in Chap. 5 assume the programs from Chap. 4 have already been run.)

5.1 Overview of the PC Algorithm

The PC algorithm is a method for constructing a graphical model DAG to infer causal relations, based on testing conditional independence among variables. It starts from the complete undirected graph, removes unnecessary edges by repeatedly testing conditional independence, and finally assigns arrow directions consistent with causal relations. The PC algorithm is particularly effective for high-dimensional data and is widely used in BN and causal inference. The name **PC**

J. Suzuki, *Graphical Models and Causal Discovery with Python*,
https://doi.org/10.1007/978-981-95-5308-2_5

algorithm (Peter–Clark Algorithm) [26] derives from its developers, Peter Spirtes and Clark Glymour. They proposed it as part of a research program in statistical causal discovery.

The PC algorithm assumes that the conditional independences among the given random variables are faithful to some graph. Below, using disjoint subsets S, T, U of $\{1, \ldots, p\}$, we write the conditional independence among random variables $X_1, \ldots, X_p$ as $X_S \perp\!\!\!\perp X_T \mid X_U$, and for (d-)separation in a graph G we write $S \perp\!\!\!\perp_G T \mid U$. Note that such triples (S, T, U) are generally not unique. We denote the conjunction over all such triples by

$$\bigwedge\nolimits_{(S,T,U)} (X_S \perp\!\!\!\perp X_T \mid X_U), \quad \bigwedge\nolimits_{(S,T,U)} (S \perp\!\!\!\perp_G T \mid U).$$

As pointed out in Chap. 3, whether G is an undirected graph (MN) or a DAG (BN), we must have

$$\bigwedge\nolimits_{(S,T,U)} (S \perp\!\!\!\perp_G T \mid U) \Longrightarrow \bigwedge\nolimits_{(S,T,U)} (X_S \perp\!\!\!\perp X_T \mid X_U), \tag{5.1}$$

but the converse

$$\bigwedge\nolimits_{(S,T,U)} (S \perp\!\!\!\perp_G T \mid U) \Longleftarrow \bigwedge\nolimits_{(S,T,U)} (X_S \perp\!\!\!\perp X_T \mid X_U) \tag{5.2}$$

does not always hold (Examples 3.9 and 3.10). Whether (5.2) holds depends on the random variables $X_1, \ldots, X_p$. If there exists a G for which both (5.1) and (5.2) hold, then $X_1, \ldots, X_p$ are said to be **faithful** to G. More precisely, we say they are faithful to an undirected graph when G is undirected and faithful to a DAG when G is a DAG.

Example 5.1 Roll dice A and dice B. Let $X_1 \in \{1, 2, 3, 4, 5, 6\}$ be the face on A; set $X_2 = 1$ if B shows an odd number and $X_2 = 0$ otherwise; and set $X_3 = 0$ if B shows at most 3 and $X_3 = 1$ if at least 4. Figure 3.6b is a BN for such X_1, X_2, X_3. Indeed, we have the conditional independences $X_1 \perp\!\!\!\perp X_2$, $X_1 \perp\!\!\!\perp X_3$, and $X_1 \perp\!\!\!\perp \{X_2, X_3\}$. Since $1 \perp\!\!\!\perp_G 2$, $1 \perp\!\!\!\perp_G 3$, and $1 \perp\!\!\!\perp_G \{2, 3\}$ hold, X_1, X_2, X_3 are faithful to a DAG. Similarly, Fig. 3.7b shows faithfulness to an undirected graph. However, if X_1 remains the face on A, X_2 is the face on B, and $X_3 = X_1 + X_2$, then the only nontrivial CI is $X_1 \perp\!\!\!\perp X_2$. This is representable as a BN in Fig. 3.6j, but among undirected graphs the only representation is Fig. 3.7h. Thus X_1, X_2, X_3 are faithful to a DAG but not to an undirected graph. ■

As shown in (3.5), for conditional independence we always have

$$X_S \perp\!\!\!\perp X_T \mid X_U \Longrightarrow \bigwedge\nolimits_{i \in S, j \in T} (X_i \perp\!\!\!\perp X_j \mid X_U), \tag{5.3}$$

but the converse need not hold.

Example 5.2 Let X_1, X_2 be independent symmetric ± 1 variables, and set $X_3 = X_1X_2$. Then $X_1 \perp\!\!\!\perp X_2$ and $X_1 \perp\!\!\!\perp X_3$ hold, but $X_1 \not\perp\!\!\!\perp \{X_2, X_3\}$. Indeed, if we know X_2 and X_3, we can deduce X_1. Hence

$$X_1 \perp\!\!\!\perp X_2,\ X_1 \perp\!\!\!\perp X_3 \not\Longrightarrow X_1 \perp\!\!\!\perp \{X_2, X_3\}.$$

For either an undirected graph or a DAG G,

$$1 \perp\!\!\!\perp_G \{2, 3\} \Longleftrightarrow X_1 \perp\!\!\!\perp \{X_2, X_3\}$$

and also

$$1 \perp\!\!\!\perp_G \{2, 3\} \Longleftrightarrow 1 \perp\!\!\!\perp_G 2,\quad 1 \perp\!\!\!\perp_G 3.$$

This is an example where (5.1) holds, but (5.2) does not. ■

For separation and d-separation, as shown in (3.8), we have for each (S, T, U)

$$\bigwedge\nolimits_{i\in S, j\in T} (i \perp\!\!\!\perp_G j \mid U) \Longleftrightarrow S \perp\!\!\!\perp_G T \mid U. \tag{5.4}$$

That is, once we know the truth values of each $i \perp\!\!\!\perp_G j \mid U$ for $(i, j) \in S \times T$, we know the truth value of $S \perp\!\!\!\perp_G T \mid U$.

Using this fact, we obtain the following proposition, central to the PC algorithm.

Proposition 5.1 *If $X_1, \ldots, X_p$ are faithful to some undirected graph or DAG, then for conditional independence*

$$\bigwedge\nolimits_{i\in S, j\in T} (X_i \perp\!\!\!\perp X_j \mid X_U) \Longrightarrow X_S \perp\!\!\!\perp X_T \mid X_U \tag{5.5}$$

holds. In other words, knowing the truth of $X_i \perp\!\!\!\perp X_j \mid X_U$ for each $(i, j) \in S \times T$ determines the truth of $X_S \perp\!\!\!\perp X_T \mid X_U$.

Proof Assuming faithfulness (i.e., (5.2)), apply (5.2), (5.4), and (5.1) in that order,

$$\begin{aligned}\bigwedge\nolimits_{i\in S, j\in T} (X_i \perp\!\!\!\perp X_j \mid X_U) &\Longrightarrow \bigwedge\nolimits_{i\in S, j\in T} (i \perp\!\!\!\perp_G j \mid U)\\ &\Longrightarrow S \perp\!\!\!\perp_G T \mid U\\ &\Longrightarrow X_S \perp\!\!\!\perp X_T \mid X_U,\end{aligned}$$

which proves (5.5). ■

Example 5.3 Suppose (X, Y, Z) follows a trivariate normal distribution. If $X \perp\!\!\!\perp Z$ and $Y \perp\!\!\!\perp Z$ both hold, then the covariance matrix has the form

$$\begin{bmatrix} * & * & 0 \\ * & * & 0 \\ 0 & 0 & * \end{bmatrix}$$

(where "*" denotes a nonzero entry), and the joint density factorizes as

$$f_{XYZ}(x, y, z) = f_{XY}(x, y) f_Z(z)$$

(Proposition 2.1). Hence $\{X, Y\} \perp\!\!\!\perp Z$ holds. The same conclusion (5.5) holds for multivariate normals of any dimension (Problem 44). ■

Under the faithfulness assumption (essentially, (5.5)), the PC algorithm restricts CI tests to the pairwise form $X_i \perp\!\!\!\perp X_j \mid X_U$ and thereby finds structure efficiently.

That is, for each $(i, j) \in S \times T$, it suffices to ensure

$$i \perp\!\!\!\perp_G j \mid X_U \Longrightarrow X_i \perp\!\!\!\perp X_j \mid X_U.$$

Concretely, for each (i, j) we find a maximal set U for which $X_i \perp\!\!\!\perp X_j \mid X_U$ holds and then use it to orient edges. Specifically:

- Build the undirected **skeleton** by CI testing on observed data (Sect. 5.3).
- Decide edge orientations on the undirected skeleton (Sect. 5.4).

These two steps produce a BN. Some edges may remain unoriented due to Markov equivalence. In such cases, applying LiNGAM (Chap. 6) is often effective.

Implementation Plan (Python) Without relying on external causal discovery packages, we implement the PC algorithm from scratch using Chap. 4 functions

`K, tilde, HSIC_2, KCI_1, null_dist.`

For skeleton estimation, for each pair (i, j) and conditioning set S, we judge independence using KCI (when $S \neq \emptyset$) or HSIC (when $S = \emptyset$) and remove edges accordingly. For orientation, we sequentially fix arrow directions using collider-detection and cycle-prevention rules, returning a CPDAG when necessary.

5.2 CI Testing in the PC Algorithm

In Chap. 4, we considered CI tests of the form $X \perp\!\!\!\perp Y \mid Z$ (with a single conditioning variable Z). In the PC algorithm we keep X, Y as they are but need to generalize to the case where Z consists of multiple variables.

In short, in testing $X \perp\!\!\!\perp Y \mid \{Z, W\}$, suppose we have Gram matrices K_X, K_Y, K_Z, K_W. Write the Hadamard (elementwise) product of K_Z and K_W as K_{ZW}. Then a test statistic that we previously formed from K_X, K_Y, K_Z for $X \perp\!\!\!\perp Y \mid Z$ is now formed from K_X, K_Y, K_{ZW} for $X \perp\!\!\!\perp Y \mid \{Z, W\}$.

Given kernels $k^{(1)}, \ldots, k^{(p)}$ for p variables and samples $x^{(1)}, \ldots, x^{(p)} \in \mathbb{R}^n$, we obtain p Gram matrices

$$K^{(1)} = (k^{(1)}(x_s^{(1)}, x_t^{(1)}))_{s,t=1,\ldots,n}, \ldots, K^{(p)} = (k^{(p)}(x_s^{(p)}, x_t^{(p)}))_{s,t=1,\ldots,n}.$$

To test

$$X_i \perp\!\!\!\perp X_j \mid X_S,$$

we use $K^{(i)}$, $K^{(j)}$, and the Hadamard product over $K^{(h)}$ for $h \in S$ to perform the CI test. In this sense, the Python function KCI_1 from Chap. 4 can be used as is. The Hadamard product can be computed in Python as follows:

```
import numpy as np
```

```
def K_merge(S, K_list):
    n = K_list[0].shape[0]
    K = np.ones((n, n), dtype=float)  # all-ones matrix (Hadamard identity),
      not I
    for h in S:
        K = K * K_list[h]             # Hadamard (elementwise) product
    return K
```

When multiplying Gram matrices, the initial value should be the all-ones matrix (not the identity). In Python, using * between arrays performs elementwise multiplication, i.e., the Hadamard product.

Using the previously defined functions, we construct the following Python routine. The p Gram matrices $K^{(1)}, \ldots, K^{(p)}$ are held in the list K_list.

```
def CI_PC(i, j, S, K_list, alpha=0.05, lam=1e-3):
    S = list(S)
    if len(S) == 0:
        res = HSIC_2(K_list[i], K_list[j])
    else:
        K_S = K_merge(S, K_list)
        res = KCI_1(K_list[i], K_list[j], K_S, lam)

    stat = float(res['statistics'])
    eigs = np.asarray(res['eigen_values'], dtype=float)

    crit = float(null_dist(eigs, alpha=alpha)['critical'])
    return (stat < crit)
```

Running the code below yields concrete truth values for conditional independence assertions.

```
# --- Sample data --
n = 100
rng = np.random.default_rng(0)
z = rng.standard_normal(n)
w = rng.standard_normal(n)
x = z + rng.standard_normal(n)
y = w + rng.standard_normal(n)

X = np.column_stack([x, y, z, w])  # [X^{(0)}, X^{(1)}, X^{(2)}, X^{(3)}]

# RBF kernel
def k_rbf(a, b, sigma2=1.0):
    a = np.asarray(a); b = np.asarray(b)
    diff = a - b
    return np.exp(-np.dot(diff, diff) / (2.0 * sigma2))

# Gram matrices for each variable
K_list = [K(k_rbf, X[:, j]) for j in range(X.shape[1])]

# CI judgments
alpha = 0.05
lam = 1e-3
res1 = CI_PC(i=0, j=1, S=[2, 3], K_list=K_list, alpha=alpha, lam=lam)
print("x␣ci␣y␣|␣{z,w}␣?", res1)  # Expected: True

# Introduce dependence: mix x into y
y_dep = y + x
X2 = np.column_stack([x, y_dep, z, w])
K_list2 = [K(k_rbf, X2[:, j]) for j in range(X2.shape[1])]
res2 = CI_PC(i=0, j=1, S=[2, 3], K_list=K_list2, alpha=alpha, lam=lam)
print("x␣not␣ci␣y␣|␣{z,w}␣?", not res2)  # Expected: True
```

Thus the CI testing method is established. However, it is not obvious in what order to test the conditional independences among p variables so that we obtain the desired undirected graph.

5.3 Constructing the Skeleton

The following procedure implements in Python the PC algorithm for constructing an undirected graph (the skeleton) based on CI tests. Initially we place all possible edges (the complete graph) and then remove unnecessary edges according to the CI tests.

We first prepare `adj`, a matrix that records adjacency among variables, initialized to TRUE everywhere except the diagonal (which is set to FALSE to prevent self-loops). Next, we prepare a 3D array `s` (whether k belongs to some maximal separating set S for which $X_i \perp\!\!\!\perp X_j \mid X_S$ holds), initialized to FALSE.

Inside the loop, for each variable pair `i`, `j` we test conditional independence for different conditioning sets `S`. The function `CI_PC(i, j, S, K_list, alpha, lambda)` tests whether `i` and `j` are conditionally independent given `S` (or independent if `S` is empty).

When `m = 0`, we test marginal independence and delete the edge if independent. When `m > 0`, we generate all subsets of size `m` using `itertools.combinations`.

For each subset S, we test $X_i \perp\!\!\!\perp X_j \mid X_S$; if confirmed, we delete the corresponding edge.

The loop terminates once no further edges are removed.

```
import numpy as np
import itertools
```

```
def skeleton(K_list, alpha=0.05, lam=1e-3, max_m=None, verbose=False):
    p = len(K_list)
    adj = np.ones((p, p), dtype=bool)
    np.fill_diagonal(adj, False)
    s = np.zeros((p, p, p), dtype=bool)

    m = 0
    while True:
        if max_m is not None and m > max_m:
            break
        removed_any = False

        for i in range(p - 1):
            for j in range(i + 1, p):
                if not adj[i, j]:
                    continue

                # current neighbors of i (excluding i, j)
                S = [k for k in range(p) if adj[i, k] and k not in (i, j)]
                if len(S) < m:
                    continue  # try only when len(S) >= m

                if m == 0:
                    indep = CI_PC(i, j, S=[], K_list=K_list, alpha=alpha, lam
                      =lam)
                    if indep:
                        adj[i, j] = adj[j, i] = False
                        removed_any = True
                        if verbose:
                            print(f"remove edge ({i},{j}) by m=0")
                else:
                    found = False
                    for cond_set in itertools.combinations(S, m):
                        indep = CI_PC(i, j, S=cond_set, K_list=K_list,
                                      alpha=alpha, lam=lam)
                        if indep:
                            adj[i, j] = adj[j, i] = False
                            for k in cond_set:
                                s[i, j, k] = True
                                s[j, i, k] = True
                            removed_any = True
                            found = True
                            if verbose:
                                print(f"remove edge ({i},{j}) with S={
                                  cond_set}")
                            break
        if not removed_any:
            break
        m += 1

    return {'adj': adj, 's': s}
```

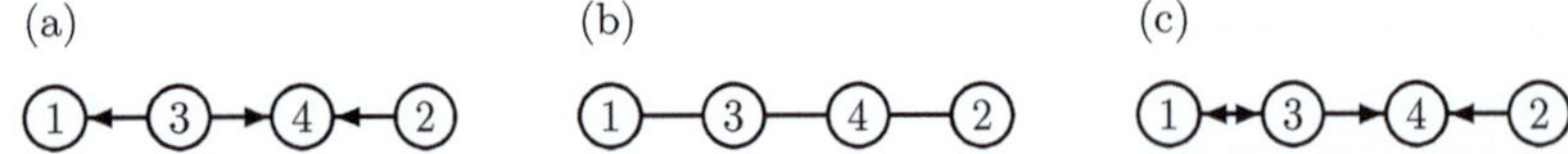

Fig. 5.1 (**a**) The true dependency structure that generated the data, (**b**) the skeleton produced by `skeleton` from the data, and (**c**) an ordering of edges on the skeleton where a collider $3 \to 4 \leftarrow 2$ is detected. The order on $1 \leftrightarrow 3$ can be chosen arbitrarily

Example 5.4 The code below generates data according to the left-hand dependency pattern in Fig. 5.1. The resulting $n \times p$ matrix is stored in `X`. From these data, we compute the skeleton using `skeleton` (Fig. 5.1, right).

```
n, p = 200, 4
rng = np.random.default_rng(0)
y = rng.standard_normal(n)
z = rng.standard_normal(n)
x = z + rng.binomial(1, 0.25, size=n).astype(float)
w = y + z + rng.binomial(1, 0.25, size=n).astype(float)
X = np.column_stack([x, y, z, w])  # shape (n, p)
# standardize
X = (X - X.mean(axis=0)) / X.std(axis=0, ddof=1)
```

The following code computes the list `K_list` of Gram matrices and estimates the skeleton.

Finally, an undirected graph is built from the adjacency matrix and drawn using `networkx`, visually displaying inter-variable dependencies.

```
# RBF kernel
def k_rbf(a, b, sigma2=1.0):
    diff = a - b
    return np.exp(- (diff * diff) / (2.0 * sigma2))
# Gram matrices for each variable
K_list = [K(k_rbf, X[:, j]) for j in range(p)]
# run skeleton estimation
alpha = 0.05
lam = 1e-3
skel = skeleton(K_list, alpha=alpha, lam=lam, verbose=True)
adj = skel['adj']    # boolean adjacency (undirected)
```

```
import networkx as nx
import matplotlib.pyplot as plt
```

```
G = nx.Graph()
for i in range(p):
    G.add_node(i, label=str(i+1))  # show 1..p

for i in range(p - 1):
    for j in range(i + 1, p):
        if adj[i, j]:
            G.add_edge(i, j)

pos = nx.spring_layout(G, seed=0)
labels = {i: G.nodes[i]['label'] for i in G.nodes}
nx.draw_networkx(G, pos=pos, labels=labels, with_labels=True)
plt.title("PC␣skeleton␣(undirected)")
plt.axis('off')
plt.show()
```

■

5.4 Orienting the Skeleton

In the second half of the PC algorithm, we orient edges on the skeleton built in the first half.

Below, we do not update the TRUE/FALSE values in `s[i,j,k]`. We only change entries in `adj[i,j]` from TRUE to FALSE to represent directions.

In Part 1 of the second half, we detect all colliders. Concretely, `skeleton` returns the array `adj`, which initially satisfies

$$\texttt{adj[i,j]} \iff \texttt{adj[j,i]}$$

(i.e., both directions are possible). Using the following rule (Rule 0), we delete one direction. Under the assumption that i and k are nonadjacent, perform the update:

$$\texttt{adj[i,j]}, \quad \texttt{adj[k,j]}, \quad \text{and not } \texttt{s[i,j,k]} \implies$$
$$\texttt{adj[j,i]} \leftarrow \text{FALSE}, \quad \texttt{adj[j,k]} \leftarrow \text{FALSE}$$

If $X_i \perp\!\!\!\perp X_j \mid X_S$ with $k \in S$, then we require `s[i,j,k] = TRUE`. A value `s[i,j,k] = FALSE` indicates a collider. Assuming that all colliders are detected in this step and that there are no others, we proceed to Part 2 and add more orientations (Fig. 5.2).

Part 2 consists of four rules.

Rule 1(a): Under the assumption that i and k are nonadjacent, if $i \to j$ and $j \leftrightarrow k$ (both directions possible between j and k), then $j \leftarrow k$ would create a collider at j. But all colliders were detected in Part 1, so this is a contradiction. Hence we conclude $j \to k$.

Rule 1(b): If $i \to j \to k$ and $i \leftrightarrow k$, then $i \leftarrow k$ would make a directed cycle; therefore we must have $i \to k$.

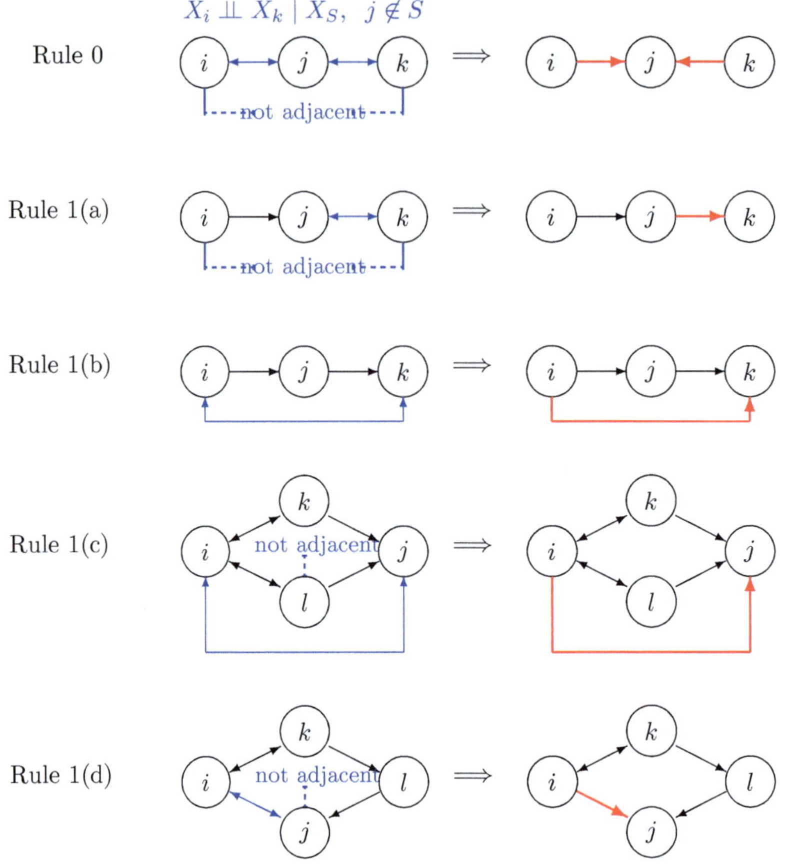

Fig. 5.2 Rule 0: When $i \perp\!\!\!\perp j \mid S(i, j)$ fails with $k \in S(i, j)$, a collider occurs. Rule 1(a): All colliders have already been detected in Rule 0, so no new collider will be introduced. Rule 1(b): Since we assume a DAG, directed cycles cannot exist. Rule 1(c): If the red arrow direction were not chosen, then no matter how we choose the three possibilities for $i \leftrightarrow k$ and $i \leftrightarrow l$ (with no collider at i), a directed cycle or a collider would arise. Rule 1(d): If $i \leftarrow j$, then either a collider arises or the directed cycle $j \to i \to k \to l \to j$ is formed

Rule 1(c): Assuming k and l are nonadjacent, if there are a collider $k \to j \leftarrow l$ and undirected edges $i \leftrightarrow j, i \leftrightarrow k, i \leftrightarrow l$, then setting $i \leftarrow j$ forces $i \leftarrow k$ and $i \leftarrow l$ to avoid cycles, which in turn introduces a new collider. Hence $i \to j$ is required.

Rule 1(d): Assuming k and j are nonadjacent, if $k \to l \to j$ and $i \leftrightarrow k, i \leftrightarrow j$, then we must set $i \to j$; otherwise, a directed cycle or a new collider appears.

With these, we implement the following Python routine. Each `adj[i,j]` entry initially has both directions TRUE (e.g., both `adj[i,j]` and `adj[j,i]`). The program represents orientation by flipping only one direction from TRUE to FALSE.

```
import numpy as np

def post_pc(adj, s, verbose=False):
    p = adj.shape[0]

    def any_(i, j):   # adjacent as an undirected edge?
        return bool(adj[i, j] or adj[j, i])

    def no(i, j):     # nonadjacent?
        return not any_(i, j)

    # ---- Rule 0: Collider detection (make j the collider) ----
    for i in range(p):
        for j in range(p):
            if j == i:
                continue
            for k in range(p):
                if k in (i, j):
                    continue
                if no(i, k) and (not s[i, k, j]) and adj[i, j] and adj[k, j]:
                    if verbose:
                        print(f"Collider at {j}: {i}-> {j} <- {k}")
                    adj[j, i] = False
                    adj[j, k] = False

    # ---- Rule 1(a): i->j, j<->k, i and k nonadjacent => j->k ----
    for i in range(p):
        for j in range(p):
            if j == i:
                continue
            for k in range(p):
                if k in (i, j):
                    continue
                if adj[i, j] and adj[j, k] and no(i, k):
                    if verbose:
                        print(f"Rule1(a): force {j}->{k}")
                    adj[k, j] = False

    # ---- Rule 1(b): i->j->k => forbid k->i (prevent cycle) ----
    for i in range(p):
        for j in range(p):
            if j == i:
                continue
            for k in range(p):
                if k in (i, j):
                    continue
                if adj[i, j] and adj[j, k]:
                    if verbose and adj[k, i]:
                        print(f"Rule1(b): forbid {k}->{i}")
                    adj[k, i] = False

    # ---- Rule 1(c): k->j<-l, and i<->j, i<->k, i<->l => i->j ----
    for j in range(p):
        for k in range(p):
            if k == j:
                continue
            for l in range(p):
                if l in (j, k):
                    continue
                if adj[k, j] and adj[l, j]:  # k->j, l->j
                    for i in range(p):
                        if i in (j, k, l):
                            continue
                        if any_(i, k) and any_(i, l) and adj[i, j]:
                            if verbose:
                                print(f"Rule1(c): force {i}->{j}")
```

```
                            adj[j, i] = False

    # ---- Rule 1(d): k->l->j, and i<->k, i<->j => i->j ----
    for j in range(p):
        for k in range(p):
            if k == j:
                continue
            for l in range(p):
                if l in (j, k):
                    continue
                if adj[k, l] and adj[l, j]:  # k->l->j
                    for i in range(p):
                        if i in (j, k, l):
                            continue
                        if any_(i, k) and adj[i, j]:
                            if verbose:
                                print(f"Rule1(d):␣force␣{i}->{j}")
                            adj[j, i] = False

    return adj
```

Example 5.5 Using `post_pc` above, we detect colliders by leveraging `s[i,j,k] = FALSE`.

```
# p = 3, edges: 1<->3, 2<->3 (internally use 0,1,2)
p = 3
adj = np.zeros((p, p), dtype=bool)
adj[0, 2] = adj[2, 0] = True  # 1 <-> 3
adj[1, 2] = adj[2, 1] = True  # 2 <-> 3
np.fill_diagonal(adj, False)

# initialize all s to True; express that 3 is NOT in the separating set for
  (1,2)
s = np.ones((p, p, p), dtype=bool)
s[0, 1, 2] = False
s[1, 0, 2] = False

adj_post = post_pc(adj.copy(), s, verbose=True)

# print as 0/1 matrix
print(adj_post.astype(int))
directed = [(i+1, j+1) for i in range(p) for j in range(p)
            if i != j and adj_post[i, j] and not adj_post[j, i]]
undirected = [(min(i,j)+1, max(i,j)+1) for i in range(p) for j in range(i+1,
  p)
              if adj_post[i, j] and adj_post[j, i]]
print("Directed:", directed)
print("Undirected:", undirected)
```

■

Example output:

```
False False True
False False True
False False False
```

Even after Part 2 completes, some edges may remain unoriented. This can be due to Markov equivalence (the same distribution is represented by multiple DAGs). If the above rules do not exhaust all constraints from "acyclic" and "no unshielded collider," it may be necessary to search for DAGs that satisfy them. Rules 1(a)–(d)

Table 5.1 Boston dataset (CRAN MASS package)

Column	Variable	Meaning of variable
1	CRIM	Per capita crime rate by town
2	ZN	Proportion of residential land zoned for lots over 25,000 sq. ft.
3	INDUS	Proportion of non-retail business acres per town
4	CHAS	Charles River dummy variable (1 if tract bounds river; 0 otherwise)
5	NOX	Nitric oxide concentration (parts per 10 million)
6	RM	Average number of rooms per dwelling
7	AGE	Proportion of owner-occupied units built prior to 1940
8	DIS	Weighted distances to five Boston employment centers
9	RAD	Index of accessibility to radial highways
10	TAX	Full-value property tax rate per $10,000
11	PTRATIO	Pupil–teacher ratio by town
12	B	Proportion of Black residents by town
13	LSTAT	Percentage of lower status population
14	MEDV	Median value of owner-occupied homes in $1000s

are representative examples that save manual effort. A mixed graph containing both directed and undirected edges in this way is sometimes called a CPDAG (completed partially directed acyclic graph).

Example 5.6 We replace the data in Example 5.4 with the housing dataset (Boston, Table 5.1) and run the procedure. HSIC and KCI rely on kernels and require roughly $O(n^3)$ time in the sample size n, so we run them on the first $n = 100$ (or $n = 200$) observations out of the full dataset (which has over 500 rows).

```
import numpy as np
import pandas as pd
```

```
# --- Load Boston data ---
df = pd.read_csv("Boston.csv")
X = df.to_numpy(dtype=float)
n = 100
X = X[:n, :]

# safe standardization (leave zero-variance columns as-is by setting std=1)
mu = X.mean(axis=0)
sd = X.std(axis=0, ddof=1)
sd_safe = sd.copy()
sd_safe[~np.isfinite(sd_safe) | (sd_safe == 0)] = 1.0  # key point
X = (X - mu) / sd_safe

def k_rbf(a, b, sigma2=1.0):
    a = np.asarray(a); b = np.asarray(b)
    diff = a - b
    return np.exp(-np.dot(diff, diff) / (2.0 * sigma2))

K_list = [K(k_rbf, X[:, j]) for j in range(X.shape[1])]
# skeleton estimation
alpha = 0.05
lam = 1e-3
```

```
out = skeleton(K_list, alpha=alpha, lam=lam, verbose=False)
adj, s = out['adj'], out['s']

# orientation
adj_dir = post_pc(adj.copy(), s, verbose=False)

# edge counts
directed_edges = [(i+1, j+1) for i in range(p) for j in range(p)
                  if i != j and adj_dir[i, j] and not adj_dir[j, i]]
undirected_edges = [(min(i,j)+1, max(i,j)+1) for i in range(p) for j in range
    (i+1, p)
                    if adj_dir[i, j] and adj_dir[j, i]]

print(f"#␣undirected␣(skeleton)␣edges:␣{sum(adj.flatten())//2}")
print(f"#␣directed␣edges␣after␣orientation:␣{len(directed_edges)}")
print("Directed␣edges␣(i->j):", directed_edges[:30], "␣...")
print("Undirected␣edges␣(i--j):", undirected_edges[:30], "␣...")
```

At the skeleton stage there are 64 directed entries (i.e., 32 undirected edges). After Rule 0, there are 33 directed entries (only $1 \leftrightarrow 11$ remains bidirectional). After applying Rules 1(a), 1(b), 1(c), and 1(d), the number of (directed) edge entries becomes 32, 30, 26, and 26, respectively (Fig. 5.3). In total, 100 TRUE instances of conditional independence were detected during skeleton estimation (Table 5.2). ■

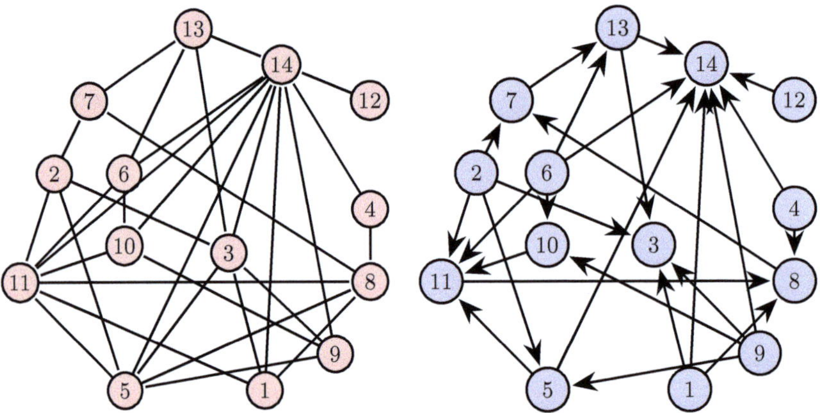

Fig. 5.3 Top: The skeleton has 32 undirected edges. After Rule 0 there are 33 directed entries; applying Rules 1(a), 1(b), 1(c), and 1(d) yields 32, 30, 26, and 26 directed entries, respectively. Bottom: the resulting CPDAG

Table 5.2 Conditional independences detected during skeleton estimation by the PC algorithm on the Boston dataset ($m \geq 4$ found none)

$m = 0$	$1 \perp\!\!\!\perp 4, 2 \perp\!\!\!\perp 4, 3 \perp\!\!\!\perp 4, 4 \perp\!\!\!\perp 5, 4 \perp\!\!\!\perp 6, 4 \perp\!\!\!\perp 7, 4 \perp\!\!\!\perp 9, 4 \perp\!\!\!\perp 10, 4 \perp\!\!\!\perp 12, 4 \perp\!\!\!\perp 13$
$m = 1$	$1 \perp\!\!\!\perp 6 \mid 5, 1 \perp\!\!\!\perp 7 \mid 5, 1 \perp\!\!\!\perp 7 \mid 8, 1 \perp\!\!\!\perp 12 \mid 9, 1 \perp\!\!\!\perp 12 \mid 10, 2 \perp\!\!\!\perp 12 \mid 1, 2 \perp\!\!\!\perp 12 \mid 3, 2 \perp\!\!\!\perp 14 \mid 3, 4 \perp\!\!\!\perp 11 \mid 14,$
	$5 \perp\!\!\!\perp 12 \mid 1, 5 \perp\!\!\!\perp 12 \mid 3, 6 \perp\!\!\!\perp 9 \mid 2, 6 \perp\!\!\!\perp 9 \mid 3, 6 \perp\!\!\!\perp 9 \mid 5, 6 \perp\!\!\!\perp 12 \mid 2, 6 \perp\!\!\!\perp 12 \mid 3, 6 \perp\!\!\!\perp 12 \mid 5, 6 \perp\!\!\!\perp 12 \mid 10,$
	$6 \perp\!\!\!\perp 12 \mid 11, 7 \perp\!\!\!\perp 9 \mid 5, 7 \perp\!\!\!\perp 9 \mid 8, 7 \perp\!\!\!\perp 10 \mid 5, 7 \perp\!\!\!\perp 10 \mid 8, 7 \perp\!\!\!\perp 11 \mid 5, 7 \perp\!\!\!\perp 12 \mid 3, 7 \perp\!\!\!\perp 12 \mid 5, 7 \perp\!\!\!\perp 12 \mid 8,$
	$8 \perp\!\!\!\perp 12 \mid 5, 9 \perp\!\!\!\perp 12 \mid 1, 9 \perp\!\!\!\perp 12 \mid 10, 9 \perp\!\!\!\perp 13 \mid 10, 10 \perp\!\!\!\perp 12 \mid 1, 11 \perp\!\!\!\perp 12 \mid 1, 11 \perp\!\!\!\perp 12 \mid 14, 11 \perp\!\!\!\perp 13 \mid 14,$
	$12 \perp\!\!\!\perp 13 \mid 14$
$m = 2$	$1 \perp\!\!\!\perp 5 \mid \{3, 9\}, 1 \perp\!\!\!\perp 5 \mid \{3, 10\}, 1 \perp\!\!\!\perp 5 \mid \{3, 11\}, 1 \perp\!\!\!\perp 5 \mid \{8, 9\}, 1 \perp\!\!\!\perp 5 \mid \{8, 10\}, 1 \perp\!\!\!\perp 5 \mid \{10, 11\},$
	$1 \perp\!\!\!\perp 6 \mid \{8, 9\}, 1 \perp\!\!\!\perp 6 \mid \{8, 10\}, 1 \perp\!\!\!\perp 6 \mid \{8, 11\}, 1 \perp\!\!\!\perp 6 \mid \{9, 13\}, 1 \perp\!\!\!\perp 6 \mid \{10, 13\}, 1 \perp\!\!\!\perp 9 \mid \{2, 10\},$
	$1 \perp\!\!\!\perp 9 \mid \{3, 10\}, 1 \perp\!\!\!\perp 9 \mid \{8, 10\}, 1 \perp\!\!\!\perp 13 \mid \{8, 14\}, 2 \perp\!\!\!\perp 6 \mid \{9, 13\}, 2 \perp\!\!\!\perp 8 \mid \{5, 9\}, 2 \perp\!\!\!\perp 9 \mid \{1, 3\},$
	$2 \perp\!\!\!\perp 9 \mid \{3, 10\}, 2 \perp\!\!\!\perp 10 \mid \{1, 11\}, 2 \perp\!\!\!\perp 10 \mid \{5, 11\}, 2 \perp\!\!\!\perp 10 \mid \{7, 11\}, 3 \perp\!\!\!\perp 6 \mid \{5, 14\}, 3 \perp\!\!\!\perp 6 \mid \{9, 14\},$
	$3 \perp\!\!\!\perp 7 \mid \{2, 5\}, 3 \perp\!\!\!\perp 7 \mid \{2, 8\}, 3 \perp\!\!\!\perp 7 \mid \{8, 14\}, 5 \perp\!\!\!\perp 13 \mid \{8, 14\}, 8 \perp\!\!\!\perp 9 \mid \{1, 7\}, 8 \perp\!\!\!\perp 9 \mid \{1, 10\},$
	$8 \perp\!\!\!\perp 9 \mid \{1, 11\}, 8 \perp\!\!\!\perp 9 \mid \{1, 13\}, 8 \perp\!\!\!\perp 9 \mid \{3, 10\}, 8 \perp\!\!\!\perp 9 \mid \{4, 5\}, 8 \perp\!\!\!\perp 9 \mid \{4, 10\}, 8 \perp\!\!\!\perp 9 \mid \{5, 6\},$
	$8 \perp\!\!\!\perp 9 \mid \{5, 10\}, 8 \perp\!\!\!\perp 9 \mid \{5, 11\}, 8 \perp\!\!\!\perp 9 \mid \{5, 13\}, 8 \perp\!\!\!\perp 9 \mid \{6, 10\}, 8 \perp\!\!\!\perp 9 \mid \{7, 10\}, 8 \perp\!\!\!\perp 9 \mid \{10, 13\},$
	$9 \perp\!\!\!\perp 11 \mid \{10, 14\}, 10 \perp\!\!\!\perp 13 \mid \{3, 9\}, 10 \perp\!\!\!\perp 13 \mid \{3, 11\}, 10 \perp\!\!\!\perp 13 \mid \{5, 14\}, 10 \perp\!\!\!\perp 13 \mid \{8, 14\},$
	$10 \perp\!\!\!\perp 13 \mid \{9, 14\}$
$m = 3$	$1 \perp\!\!\!\perp 2 \mid \{8, 11, 14\}, 3 \perp\!\!\!\perp 11 \mid \{1, 5, 14\}, 3 \perp\!\!\!\perp 11 \mid \{1, 8, 14\}, 3 \perp\!\!\!\perp 11 \mid \{2, 5, 14\}, 5 \perp\!\!\!\perp 10 \mid \{3, 7, 9\},$
	$5 \perp\!\!\!\perp 10 \mid \{3, 8, 9\}, 6 \perp\!\!\!\perp 8 \mid \{10, 13, 14\}$

Problems 44–53

44. In Example 5.3, consider the p-variate normal case. Let $\mathcal{X}, \mathcal{Y}, \mathcal{Z}$ be disjoint subsets of $\{1, \ldots, p\}$ with sizes s, t, u, respectively, and $p = s + t + u$. When both $\mathcal{X} \perp\!\!\!\perp \mathcal{Z}$ and $\mathcal{Y} \perp\!\!\!\perp \mathcal{Z}$ hold, how many of the p^2 entries of the covariance matrix are zero?
45. When the following program is executed, what properties does the resulting data `gmG` have?

```
import numpy as np, networkx as nx
```

```
def random_DAG(p, prob=0.5, V=None):
    if V is None:
        V = [str(i+1) for i in range(p)]
    G = nx.DiGraph()
    G.add_nodes_from(V)
    for i in range(p):
        for j in range(i+1, p):
            if np.random.rand() < prob:
                G.add_edge(V[i], V[j])
    return G
```

```
def rmv_DAG(n, G):
    nodes = list(nx.topological_sort(G))
    p = len(nodes)
    B = np.zeros((p, p))
    for i,u in enumerate(nodes):
        for j,v in enumerate(nodes):
            if G.has_edge(u,v):
                B[j,i] = np.random.uniform(0.5,1.0)
    X = np.zeros((n,p))
    for t in range(n):
        eps = np.random.normal(size=p)
        for j in range(p):
            X[t,j] = np.dot(B[j,:j], X[t,:j]) + eps[j]
    return X
```

```
p, n = 10, 1000
vars = ["1"] + [str(i) for i in range(2,p+1)]
gGtrue = random_DAG(p, prob=0.5, V=vars)
gmG = {"x": rmv_DAG(n, gGtrue), "g": gGtrue}
```

46. Using the data `gmG` obtained above, run the PC algorithm, and compute the CPDAG by using the functions defined in Chaps. 3 and 4.
47. In the function `K_merge`, why must the initial value of the matrix `K` *not* be the identity matrix?
48. In the function `skeleton`, explain why we process only when `len(S) >= m`. Also, under what condition is `s[i,j,k]` set to TRUE?
49. Why must i and k be nonadjacent in Rule 0 and Rule 1(a)?
50. For the 11 DAGs in Fig. 3.6, after relabeling X, Y, Z as 1, 2, 3, respectively, determine TRUE/FALSE for $S[i, j, k]$ for each $(i, j, k) \in \{(1, 2, 3), (1, 3, 2), (2, 1, 3), (2, 3, 1), (3, 1, 2), (3, 2, 1)\}$.
51. Given the undirected graph

$$1 - 3, \quad 3 - 2, \quad 2 - 4, \quad 4 - 1$$

and $p = 4$ with $s[1, 2, 3] = s[2, 1, 3] =$ FALSE, $s[1, 2, 4] = s[2, 1, 4] =$ TRUE, and all other entries TRUE, show the orientation result of `Post_PC(adj, s)`. Explain which rule orients which edge.

52. In Example 5.6, draw the CPDAG obtained immediately after applying Rule 0 and then after applying Rules 1(a), 1(b), 1(c), and 1(d), respectively.
53. Given vectors x, y (length n), a matrix z ($n \times p$), and significance level α, implement a function that tests $X \perp\!\!\!\perp Y \mid Z_1, \ldots, Z_p$. For $n = 100$, $p = 2$, generate random data for a case where the CI holds and a case where it does not, and perform the tests.

Chapter 6
LiNGAM

In this chapter we present LiNGAM (Linear Non-Gaussian Acyclic Model) [21, 22], a method that identifies a causal ordering of variables based on non-Gaussianity. In contrast to the PC algorithm and score-based structure learning, LiNGAM assumes an additive noise model and directly recovers a causal order of variables.

We first introduce Direct LiNGAM, which assumes an ordering in a linear structural model, computes residuals, and detects upstream variables. As theoretical background for identifiability of the order, we explain the Darmois–Skitovich theorem [5, 25] in an accessible way. We then describe ICA LiNGAM [21], which is based on independent component analysis (ICA), together with the mathematical background and the algorithmic flow.

We also touch on extensions of LiNGAM under latent confounding, including theoretical results by Wang–Drton and a shortest-path-based reconstruction method, thereby highlighting the potential of causal discovery leveraging non-Gaussianity.

(The programs in Chap. 6 assume that the programs from Chaps. 4 and 5 have already been executed.)

6.1 Direct LiNGAM

LiNGAM (Linear Non-Gaussian Model) is a causal discovery method devised by Shohei Shimizu. While several variants exist, in this chapter we focus on ICA LiNGAM (2006) [21] and Direct LiNGAM (2011) [22]. We begin with Direct LiNGAM (Fig. 6.1).

Suppose random variables X, Y can be written either as

$$\begin{cases} X = e_1 \\ Y = aX + e_2 \end{cases} \tag{6.1}$$

J. Suzuki, *Graphical Models and Causal Discovery with Python*,
https://doi.org/10.1007/978-981-95-5308-2_6

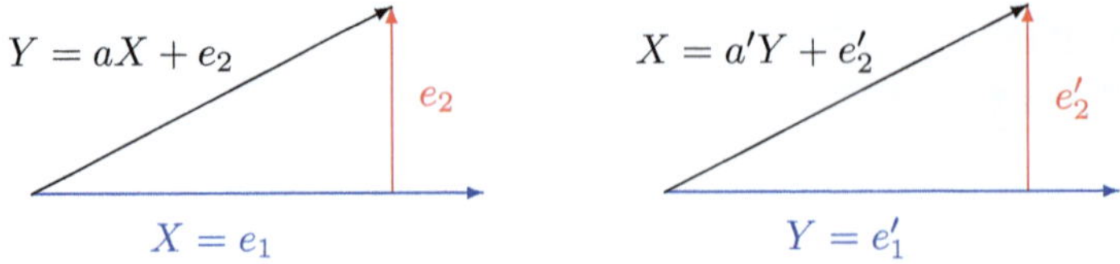

Fig. 6.1 In model (6.1), X is generated independently of Y and Y adds noise e_2 independent of $X = e_1$. Model (6.2) is the reverse

for some constant $a \in \mathbb{R}$ and mutually independent, mean-zero noises e_1, e_2 or as

$$\begin{cases} Y = e_1' \\ X = a'Y + e_2' \end{cases} \tag{6.2}$$

for some constant $a' \in \mathbb{R}$ and mutually independent, mean-zero noises e_1', e_2'. If the former holds, we say X causes Y and write $X \to Y$; if the latter holds, we write $Y \to X$. This is called an **additive noise model** [10].

Model (6.1) states that X is generated independently of Y and Y equals a linear multiple of X plus noise e_2 independent of $X = e_1$. Conversely, (6.2) states that Y is generated independently of X and X equals a linear multiple of Y plus noise e_2' independent of $Y = e_1'$.

Here the constants a, a' are chosen to minimize mean squared error. Differentiating

$$\mathbb{E}_{XY}[\|Y - aX\|^2]$$

with respect to a and setting to zero yield

$$-2\,\mathbb{E}_{XY}[X(Y - aX)] = 0, \tag{6.3}$$

so as long as X is not almost surely zero,

$$a = \frac{\mathbb{E}_{XY}[XY]}{\mathbb{E}_X[X^2]}. \tag{6.4}$$

Under (6.3) and (6.4), the covariance is

$$\text{cov}(e_1, e_2) = \mathbb{E}_{XY}[X(Y - aX)] = 0. \tag{6.5}$$

The same argument for a' gives $\text{cov}(e_1', e_2') = 0$. Recall that if (X, Y) are jointly normal, $\text{cov}(X, Y) = 0$ implies $X \perp\!\!\!\perp Y$ (Chap. 2, Proposition 2). Therefore unless at least one of the noise pairs fails to be independent, both pairs become independent ($e_1 \perp\!\!\!\perp e_2$ and $e_1' \perp\!\!\!\perp e_2'$), and the direction is not identifiable. We study identifiability in Sect. 6.3.

In practice, given i.i.d. samples of X, Y, we decide whether (6.1) or (6.2) better fits. For

$$x^n := (x_1, \dots, x_n), \qquad y^n := (y_1, \dots, y_n) \in \mathbb{R}^n,$$

center by subtracting $\bar{x} := \frac{1}{n}\sum_{i=1}^n x_i$ and $\bar{y} := \frac{1}{n}\sum_{i=1}^n y_i$. Define empirical variances/covariance

$$v(x^n) := \frac{1}{n}\sum_{i=1}^n x_i^2, \qquad v(y^n) := \frac{1}{n}\sum_{i=1}^n y_i^2, \qquad c(x^n, y^n) := \frac{1}{n}\sum_{i=1}^n x_i y_i$$

and residuals

$$y_x^n := y^n - x^n \cdot \frac{c(x^n, y^n)}{v(x^n)}, \qquad x_y^n := x^n - y^n \cdot \frac{c(x^n, y^n)}{v(y^n)}. \tag{6.6}$$

We then assess which is "closer to independence":

$$x^n \perp\!\!\!\perp y_x^n \qquad \text{vs.} \qquad y^n \perp\!\!\!\perp x_y^n.$$

One may compare the estimates of mutual information

$$I(X, Y) := \int_{\mathcal{X}}\int_{\mathcal{Y}} f_{XY}(x, y) \log \frac{f_{XY}(x, y)}{f_X(x) f_Y(y)}\, dx\, dy,$$

which equals zero iff $X \perp\!\!\!\perp Y$ (Sect. 3.4). However, LiNGAM commonly uses HSIC introduced in Chap. 4:

1. Compare the HSIC estimator (4.7) and choose the smaller value.
2. Compare the p-values from HSIC-based independence tests, and choose the larger p-value.

Absolute HSIC values are not comparable across unrelated settings, so method (1) lacks a firm theoretical justification, though smaller values provide evidence of near-independence. Method (2) is more computationally intensive but statistically more principled.

First we implement method (1) in Python. The function `LiNGAM_1(x,y)` evaluates the direction $X \to Y$; the smaller the HSIC, the more plausible $X \to Y$.

```
import numpy as np
from sklearn.decomposition import FastICA
```

```
def LiNGAM_1(x, y, proc="HSIC"):
    sigma2 = 1.0
    def k_x(a, b): return np.exp(-np.linalg.norm(a - b) ** 2 / (2 * sigma2))
    k_z = k_x
    z = y - np.sum(x * y) / np.sum(x ** 2) * x
    K_x = K(k_x, x)
    K_z = K(k_z, z)
    result = HSIC_2(K_x, K_z)
    if proc == "HSIC":
        return result["statistics"]
    else:
        seq = null_dist(result["eigen_values"], 0.05)["z"]
        return np.mean(result["statistics"] < seq)
```

Example 6.1 Compare `LiNGAM_1(x, y)` with `LiNGAM_1(y, x)`. If the former is smaller, infer $X \to Y$; if the latter is smaller, infer $Y \to X$. Before running, define from Chap. 3 the kernel Gram matrix function `K`, the centering function `tilde`, the HSIC function `HSIC_2`, and the null-distribution builder `null_dist`. Using `LiNGAM_1`, compute HSIC for (x^n, y^n) (testing $X \to Y$), HSIC for (y^n, x^n) (testing $Y \to X$), and their empirical p-values.

```
n = 100
x = np.random.randn(n)
y = x + (np.random.randn(n) ** 2 - np.mean(np.random.randn(n) ** 2))
print(LiNGAM_1(x, y, proc="HSIC"))
print(LiNGAM_1(y, x, proc="HSIC"))
print(LiNGAM_1(x, y, proc="p.val"))
print(LiNGAM_1(y, x, proc="p.val"))
```

In this example $X \to Y$ is true; `LiNGAM_1(x,y)` is smaller, so the direction is correctly estimated. ■

6.2 Extension to Multivariate Cases

Model (6.1) extends to three variables X, Y, Z as follows: For some $a, b, c \in \mathbb{R}$ and mutually independent e_1, e_2, e_3,

$$\begin{cases} X = e_1 \\ Y = aX + e_2 \\ Z = bX + cY + e_3 \end{cases} \tag{6.7}$$

in which case we write $X \to Y \to Z$; the other five orderings are analogous. Set

$$a = \frac{\mathbb{E}_{XY}[XY]}{\mathbb{E}_X[X^2]}, \qquad b = \frac{\mathbb{E}_{XZ}[XZ]}{\mathbb{E}_X[X^2]}, \qquad c = \frac{\mathbb{E}_{YZ}[YZ]}{\mathbb{E}_Y[Y^2]},$$

$$a' = \frac{\mathbb{E}_{XY}[XY]}{\mathbb{E}_Y[Y^2]}, \qquad b' = \frac{\mathbb{E}_{XZ}[XZ]}{\mathbb{E}_Z[Z^2]}, \qquad c' = \frac{\mathbb{E}_{YZ}[YZ]}{\mathbb{E}_Z[Z^2]}.$$

Compare

$$X \perp\!\!\!\perp \{Y - aX,\ Z - bX\}, \quad Y \perp\!\!\!\perp \{Z - cY,\ X - a'Y\}, \quad Z \perp\!\!\!\perp \{Xvb'Z,\ Y - c'Z\}.$$

That is, identify the most upstream variable, and test its independence from the two residuals. If the first condition is judged true (i.e., X is the most upstream), then compare

$$(Y - aX) \perp\!\!\!\perp \left((Z - bX) - (Y - aX) \cdot \frac{\mathbb{E}_{XYZ}[(Z - bX)(Y - aX)]}{\mathbb{E}_{XY}[(Y - aX)^2]}\right),$$

$$(Z - bX) \perp\!\!\!\perp \left((Y - aX) - (Z - bX) \cdot \frac{\mathbb{E}_{XYZ}[(Y - aX)(Z - bX)]}{\mathbb{E}_{XZ}[(Z - bX)^2]}\right).$$

If the former holds, infer $X \to Y \to Z$; if the latter holds, infer $X \to Z \to Y$.

Concretely, for samples

$$x^n = (x_1, \ldots, x_n), \quad y^n = (y_1, \ldots, y_n), \quad z^n = (z_1, \ldots, z_n),$$

first determine which of

$$x^n \perp\!\!\!\perp \{y_x^n, z_x^n\}, \qquad y^n \perp\!\!\!\perp \{z_y^n, x_y^n\}, \qquad z^n \perp\!\!\!\perp \{x_z^n, y_z^n\}$$

holds (with $x_z^n, y_z^n, z_x^n, z_y^n$ defined as in (6.6)). If, for example, $x^n \perp\!\!\!\perp \{y_x^n, z_x^n\}$ is true, define residuals

$$y_{xz}^n := y_x^n - \frac{c(y_x^n, z_x^n)}{v(z_x^n)} z_x^n, \qquad z_{xy}^n := z_x^n - \frac{c(y_x^n, z_x^n)}{v(y_x^n)} y_x^n, \tag{6.8}$$

and decide which of

$$y_x^n \perp\!\!\!\perp z_{xy}^n \qquad \text{vs.} \qquad z_x^n \perp\!\!\!\perp y_{xz}^n$$

holds. The former implies $X \to Y \to Z$. Overall we obtain the following decision flow:

$$\begin{cases} x^n \perp\!\!\!\perp \{y_x^n, z_x^n\} \Longrightarrow \begin{cases} y_x^n \perp\!\!\!\perp z_{xy}^n \Longrightarrow X \to Y \to Z \\ z_x^n \perp\!\!\!\perp y_{xz}^n \Longrightarrow X \to Z \to Y \end{cases} \\ y^n \perp\!\!\!\perp \{z_y^n, x_y^n\} \Longrightarrow \begin{cases} z_y^n \perp\!\!\!\perp x_{yz}^n \Longrightarrow Y \to Z \to X \\ x_y^n \perp\!\!\!\perp z_{yx}^n \Longrightarrow X \to Z \to Y \end{cases} \\ z^n \perp\!\!\!\perp \{x_z^n, y_z^n\} \Longrightarrow \begin{cases} x_z^n \perp\!\!\!\perp y_{zx}^n \Longrightarrow Z \to X \to Y \\ y_z^n \perp\!\!\!\perp x_{zy}^n \Longrightarrow Z \to Y \to X. \end{cases} \end{cases}$$

Moreover,

$$x^n_{yz} = x^n_{zy}, \qquad y^n_{zx} = y^n_{xz}, \qquad z^n_{xy} = z^n_{yx}$$

holds (Fig. 6.2). In fact, z^n_{xy} can be written in terms of x^n, y^n, z^n as (see the Appendix for a proof)

$$z^n - \frac{v(y^n)c(x^n, z^n) - c(x^n, y^n)c(y^n, z^n)}{v(x^n)v(y^n) - c(x^n, y^n)^2} x^n - \frac{v(x^n)c(y^n, z^n) - c(x^n, y^n)c(x^n, z^n)}{v(x^n)v(y^n) - c(x^n, y^n)^2} y^n. \tag{6.9}$$

Swapping x^n and y^n interchanges the second and third terms, leaving the whole expression unchanged. Without proof, the same order-independence of subscripts holds for $p \geq 4$ variables (Fig. 6.2).

The multivariate case still seeks an ordering that makes residuals independent. We extend the implementation to handle $p \geq 3$ (Fig. 6.3):

```
def LiNGAM_2(i, Z, proc="HSIC"):
    """
␣␣␣␣Evaluate␣whether␣variable␣i␣is␣most␣upstream␣given␣list␣Z␣=␣[Z_0,...,Z_{m
  -1}].
␣␣␣␣Returns␣HSIC␣(or␣p-value)␣between␣X=Z[i]␣and␣the␣Hadamard-product␣kernel
␣␣␣␣of␣residualized␣others.
␣␣␣␣"""
    sigma2 = 1.0

    # Gaussian kernel (corresponds to R's k.x)
    def k_x(a, b):
        return np.exp(-np.linalg.norm(a - b) ** 2 / (2.0 * sigma2))

    # Sizes
```

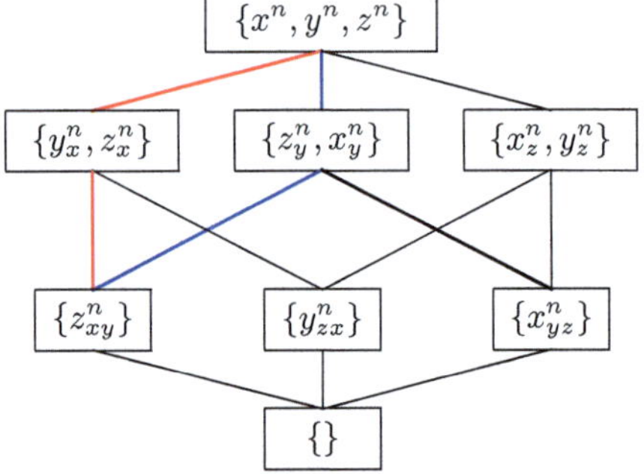

Fig. 6.2 Residuals of upstream variables propagate downstream. The paths $X \to Y \to Z$ (shown in red) and $Y \to X \to Z$ (shown in blue) merge at $\{z^n_{xy}\}$, but even though the order of X and Y differs, the residuals are equal: $z_{xy} = z_{yx}$

$$X \to Y \to Z: \quad \{x^n, y^n, z^n\} \to \{y^n_x, z^n_x\} \to \{z^n_{xy}\} \to \{\}$$
$$Y \to X \to Z: \quad \{x^n, y^n, z^n\} \to \{z^n_y, x^n_y\} \to \{z^n_{yx}\} \to \{\}$$
$$Y \to Z \to X: \quad \{x^n, y^n, z^n\} \to \{z^n_y, x^n_y\} \to \{x^n_{yz}\} \to \{\}$$
$$X \to Z \to Y: \quad \{x^n, y^n, z^n\} \to \{y^n_z, z^n_x\} \to \{y^n_{xz}\} \to \{\}$$
$$Z \to X \to Y: \quad \{x^n, y^n, z^n\} \to \{y^n_z, x^n_z\} \to \{y^n_{zx}\} \to \{\}$$
$$Z \to Y \to X: \quad \{x^n, y^n, z^n\} \to \{x^n_z, y^n_z\} \to \{x^n_{zy}\} \to \{\}$$

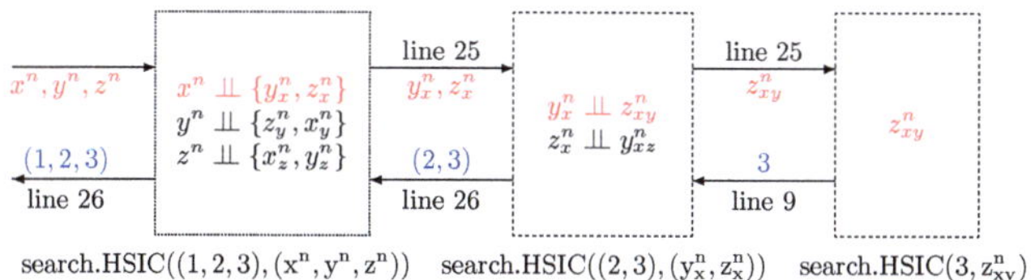

Fig. 6.3 When computing search.HSIC((1, 2, 3), (x^n, y^n, z^n)), if variable 1 is judged to be the most upstream, at line 25 it requests the computation of search.HSIC((2, 3), (y^n_x, z^n_x)). In the next process to the right, if variable 2 is judged to be the next upstream, then at line 25 it requests the computation of search.HSIC(3, z^n_{xy}). However, in the process on the right, since there is nothing further to compare, it returns the value 3 from line 9. In the process to the left, it returns the value (2,3) from line 26. Finally, the leftmost process returns the value (1,2,3) as the final output

```
    n = len(Z[i])
    m = len(Z)

    # Kernel for X side
    K_x = K(k_x, Z[i])

    # Kernel for residual side, start from all-ones (Hadamard identity)
    K_z = np.ones((n, n))

    # For each j != i, regress Z[j] on Z[i], take residual, and accumulate
      kernels
    idx = [j for j in range(m) if j != i]
    Zi = Z[i]
    ZiZi = np.sum(Zi * Zi)  # sum(z_i^2)
    Z_new = Z[:]            # local copy (caller should pass Z.copy() if
      needed)

    for j in idx:
        coef = np.sum(Zi * Z_new[j]) / ZiZi
        Z_new[j] = Z_new[j] - coef * Zi     # residual
        K_z = K_z * K(k_x, Z_new[j])        # Hadamard product

    # HSIC
    result = HSIC_2(K_x, K_z)

    # Pack residual list (excluding i)
    Z_minus_i = [Z_new[j] for j in idx]
    if proc == "HSIC":
        return {"HSIC": result["statistics"], "Z": Z_minus_i}
    else:
        # Empirical null via eigenvalues, one-sided p-value
        seq = null_dist(result["eigen_values"], 0.05)["z"]
        p_val = np.mean(result["statistics"] < seq)
        return {"p.val": p_val, "Z": Z_minus_i}
```

Using `LiNGAM_2`, we search an order by HSIC:

```
def search_HSIC(Z, index=None):
    """
    Given Z = [z_0, ..., z_{m-1}] (each a 1D array),
    recursively find a causal order minimizing HSIC at each step.
    """
    m = len(Z)
    if index is None:
```

```
        index = list(range(m))
    if m < 2:
        return index

    min_val = np.inf
    for i in range(m):
        result = LiNGAM_2(i, Z.copy(), proc="HSIC")
        if result["HSIC"] < min_val:
            min_val = result["HSIC"]
            W = result["Z"]
            k = index[i]

    index_3 = search_HSIC(W, [j for j in index if j != k])
    return [k] + index_3
```

Similarly, one can define search_pval that uses p-values instead of HSIC (Problem 55).

Example 6.2 We tested direction recovery on the following synthetic data. While a single run cannot guarantee performance, search_pval—although slower—often yields more accurate orders.

```
n = 100
x = (np.random.randn(n) ** 2 - np.mean(np.random.randn(n) ** 2))
y = 2 * x + (np.random.randn(n) ** 2 - np.mean(np.random.randn(n) ** 2))
z = -3 * x + 4 * y + (np.random.randn(n) ** 2 - np.mean(np.random.randn(n) **
    2))
X = [x, y, z]
print(search_HSIC(X))  # expected: 0 1 2
print(search_pval(X))  # expected: 0 1 2
```

In this experiment, search_HSIC and search_pval return the same order. In other settings they may disagree, and the p-value version can be correct when the HSIC-only version is not. ■

Finally, we apply LiNGAM-based order search to real data.

Example 6.3 Apply search_HSIC to the Boston dataset. The following order was obtained:

```
# Boston dataset
df = pd.read_csv("Boston.csv")
X = df.to_numpy(dtype=float)
n = 100

# Take rows 100:200 and only center (mean 0)
B_sub = X[100:100+n, :]
B_centered = B_sub - B_sub.mean(axis=0, keepdims=True)

# Feed each column to search_HSIC
Z = [B_centered[:, j] for j in range(B_centered.shape[1])]
order = search_HSIC(Z)

# Shift to 1-based for display
print([i+1 for i in order])
```

Example output:

```
[5, 4, 6, 1, 8, 9, 11, 13, 14, 3, 2, 7, 10, 12]
```

This order may differ from that obtained by the PC algorithm in Chap. 5. The PC algorithm constructs a DAG solely from conditional independence, whereas LiNGAM imposes an additive noise model and imposes an order even where PC leaves equivalences. ■

If we repeat Example 6.3 with `search_pval(Boston)`, all p-values may become 0 and no order is returned. LiNGAM assumes independent noises; in (6.1)–(6.2), at least one of $e_1 \perp\!\!\!\perp e_2$ or $e'_1 \perp\!\!\!\perp e'_2$ must hold. In real data this is often violated (e.g., due to confounding). We return to this in Sect. 5.4.

6.3 Identifiability

Thus far we discussed procedures. We now study theoretical properties.

A natural question is if both (6.1) and (6.2) hold simultaneously, can we say $X \to Y$ or $Y \to X$? LiNGAM postulates that at most one holds. If exactly one holds, the direction is **identifiable**; if both hold, it is not.

Example 6.4 Let (6.1) hold with $a = 1$ and independent $e_1, e_2 \sim N(0, 1)$. For any $a' \in \mathbb{R}$ define

$$\begin{cases} e'_1 = Y = e_1 + e_2, \\ e'_2 = X - a'Y = (1 - a')e_1 - a'e_2. \end{cases}$$

The covariance of e'_1, e'_2 is $1 - 2a'$, which is zero for $a' = 1/2$. Since linear combinations of normals are normal (Proposition 2.3), both e'_1, e'_2 are normal; by Proposition 2.1 they are independent. Hence both (6.1) and (6.2) hold simultaneously. ■

We next derive general conditions under which $e_1 \perp\!\!\!\perp e_2$ and $e'_1 \perp\!\!\!\perp e'_2$ cannot both hold. If both representations exist, then

$$\begin{cases} X = e_1 \\ Y = aX + e_2 \end{cases} \quad \text{and} \quad \begin{cases} Y = e'_1 \\ X = a'Y + e'_2 \end{cases} \Longrightarrow \begin{cases} e'_1 = ae_1 + e_2 \\ e_1 = a'e'_1 + e'_2 \end{cases}$$

and thus

$$e'_2 = e_1 - a'e'_1 = (1 - aa')e_1 - a'e_2,$$

i.e.,

$$\begin{cases} e'_1 = ae_1 + e_2 \\ e'_2 = (1 - aa')e_1 - a'e_2. \end{cases} \tag{6.10}$$

If $e_1 \perp\!\!\!\perp e_2$ with variances σ_1^2, σ_2^2, then $\mathrm{cov}(e_1', e_2') = a(1 - aa')\sigma_1^2 - a'\sigma_2^2$, which is zero for

$$a' = \frac{a\sigma_1^2}{a^2\sigma_1^2 + \sigma_2^2}. \tag{6.11}$$

Thus if e_1, e_2 are normal, for any $a \in \mathbb{R}$ there exists a' such that both $e_1 \perp\!\!\!\perp e_2$ and $e_1' \perp\!\!\!\perp e_2'$ hold.

The converse is also true. Using the following classical fact, we see that both $e_1 \perp\!\!\!\perp e_2$ and $e_1' \perp\!\!\!\perp e_2'$ can hold only in the normal case.

Proposition 6.1 (Darmois–Skitovich, 1953) *Let $X_1, \ldots, X_m$ be mutually independent, nondegenerate random variables, and Y_1, Y_2 independent of each other satisfy*

$$Y_1 = \alpha_1 X_1 + \cdots + \alpha_m X_m, \tag{6.12}$$

$$Y_2 = \beta_1 X_1 + \cdots + \beta_m X_m, \tag{6.13}$$

for constants $\alpha_i, \beta_i \in \mathbb{R}$. Suppose there are no $i \neq j$ with $\alpha_i \beta_j - \alpha_j \beta_i = 0$.[1] *Then for every i with $\alpha_i \beta_i \neq 0$, X_i must be normally distributed.*

The proof is given in the Appendix.

Apply Proposition 6.1 to $Y_1 = e_1'$, $Y_2 = e_2'$ in (6.10) with $m = 2$. For both $e_1 \perp\!\!\!\perp e_2$ and $e_1' \perp\!\!\!\perp e_2'$ to hold, we need (6.11), so we may set

$$(\alpha_1, \alpha_2, \beta_1, \beta_2) = \left(a,\ 1,\ 1 - aa',\ -a'\right) = \left(a,\ 1,\ \frac{\sigma_2^2}{a^2\sigma_1^2 + \sigma_2^2},\ -\frac{a\sigma_1^2}{a^2\sigma_1^2 + \sigma_2^2}\right).$$

Then

$$\alpha_1 \beta_1 = \frac{a\sigma_2^2}{a^2\sigma_1^2 + \sigma_2^2}, \qquad \alpha_2 \beta_2 = -\frac{a\sigma_1^2}{a^2\sigma_1^2 + \sigma_2^2}.$$

Hence, except for $a = 0$ (equivalently $a' = 0$ and $X \perp\!\!\!\perp Y$; Problem 57), e_1 and e_2 must both be normal. We summarize the following:

Proposition 6.2 (Shimizu et al. 2011) *Assume $X \not\perp\!\!\!\perp Y$. Then the following are equivalent: (i) non-identifiability (both $e_1 \perp\!\!\!\perp e_2$ and $e_1' \perp\!\!\!\perp e_2'$), (ii) e_1, e_2 are normal, and (iii) e_1', e_2' are normal.*

We next consider extensions of Proposition 6.2 to multivariate settings.

[1] Otherwise $\alpha_i X_i + \alpha_j X_j$ is a scalar multiple of $\beta_i X_i + \beta_j X_j$ and can be treated as a single variable.

Proposition 6.3 *If at least* $p - 1$ *of the noises* $e_1, \ldots, e_p$ *are non-Gaussian, then the order of* $X_1, \ldots, X_p$ *is identifiable.*

The proof is given in the Appendix.

Proposition 6.3 is only sufficient. Even with two or more Gaussian noises, identifiability can still hold (Problems 59–60).

6.4 Independent Component Analysis (ICA)

Before discussing ICA LiNGAM, we briefly review **Independent Component Analysis (ICA)**.

In Direct LiNGAM we primarily assumed a **structural equation model** (SEM). Namely, for $i = 1, \ldots, p$,

$$X_i = \sum_{j=1}^{i-1} b_{i,j} X_j + e_i,$$

where $e_1, \ldots, e_p$ are mutually independent. Let $B = (b_{i,j})$ have nonzero entries only strictly below the diagonal, and write $X = [X_1, \ldots, X_p]^\top$, $e = [e_1, \ldots, e_p]^\top$. Then

$$\begin{bmatrix} X_1 \\ \vdots \\ X_p \end{bmatrix} = B \begin{bmatrix} X_1 \\ \vdots \\ X_p \end{bmatrix} + \begin{bmatrix} e_1 \\ \vdots \\ e_p \end{bmatrix} \quad \text{or equivalently} \quad \begin{bmatrix} X_1 \\ \vdots \\ X_p \end{bmatrix} = (I - B)^{-1} \begin{bmatrix} e_1 \\ \vdots \\ e_p \end{bmatrix}. \tag{6.14}$$

Direct LiNGAM treats B as unknown and, from n realizations of $X \in \mathbb{R}^p$, estimates the independent noises $e_1, \ldots, e_p$. More generally, given $X \in \mathbb{R}^p$, ICA estimates independent components $S = [s_1, \ldots, s_m]^\top$ with $m \le p$ such that $X = WS$.

WLOG assume that X and e have zero mean and that among the m sources, at most one s_j is Gaussian. Indeed, if $X \sim N(0, \sigma_1^2)$ and $Y \sim N(0, \sigma_2^2)$, then by *reproducibility* (Prop. 2.3) $X + Y \sim N(0, \sigma_1^2 + \sigma_2^2)$. Moreover (Cramér's theorem [17]) if X and Y are independent and $X + Y \sim N(0, \sigma_1^2 + \sigma_2^2)$, then X, Y are necessarily zero-mean Gaussian. In that case, due to reproducibility, for any positive $\tilde{\sigma}_1^2, \tilde{\sigma}_2^2$ with $\tilde{\sigma}_1^2 + \tilde{\sigma}_2^2 = \sigma_1^2 + \sigma_2^2$, we can redistribute variance, implying that Gaussian components cannot be separated from a Gaussian linear mixture.

Thus, extracting independent components amounts to extracting non-Gaussian components. Among several ICA implementations, a common idea is to extract the direction most deviant from Gaussianity, then, within the subspace orthogonal to it, extract the next-most non-Gaussian direction, and so on.

In `scikit-learn`, `FastICA` first centers and scales $X \in \mathbb{R}^{n\times p}$ and then whitens it. With sample covariance $V := \frac{1}{n}X^\top X$, eigenvectors $U = [u_1, \ldots, u_p] \in \mathbb{R}^{p\times p}$, diagonal $D = \mathrm{diag}(\lambda_1, \ldots, \lambda_p)$, and $D^{-1/2} = \mathrm{diag}(\lambda_1^{-1/2}, \ldots, \lambda_p^{-1/2})$, define $\tilde{X} := XUD^{-1/2} \in \mathbb{R}^{n\times p}$. Using $U^\top U = I$ and $VU = UD$,

$$\frac{1}{n}\tilde{X}^\top\tilde{X} = D^{-1/2}U^\top\left(\frac{1}{n}X^\top X\right)UD^{-1/2} = D^{-1/2}U^\top VUD^{-1/2} = I.$$

Thus columns of $\tilde{X}$ are uncorrelated. Below we re-denote this whitened matrix by $X \in \mathbb{R}^{n\times p}$.

For $X = (X_{i,j})$, define $s_i = \sum_{j=1}^p X_{i,j}w_j$ $(i = 1, \ldots, n)$ and, under $\sum_{j=1}^p w_j^2 = 1$, seek $w = (w_1, \ldots, w_p)$ maximizing $(1/n)\sum_{i=1}^n G(s_i)$, where G measures non-Gaussianity (see candidates in Problem 61).

Let f, g be densities, and define the Kullback–Leibler divergence

$$D(f\|g) := \mathbb{E}_X\left[\log\frac{f(X)}{g(X)}\right] = \int f(x)\log\frac{f(x)}{g(x)}\,dx.$$

It is nonnegative and equals 0 iff $f(X) = g(X)$ a.s. Let

$$f_{\mathrm{gauss}}(x) = \frac{1}{\sqrt{2\pi\sigma^2}}\exp\left(-\frac{x^2}{2\sigma^2}\right),$$

and assume

$$\int x^2 f(x)\,dx = \int x^2 f_{\mathrm{gauss}}(x)\,dx = \sigma^2. \tag{6.15}$$

Then for any density f,

$$h(f) := \int -f(x)\log f(x)\,dx \;\le\; h(f_{\mathrm{gauss}}) = \frac{1}{2}\log(2\pi e\,\sigma^2) \tag{6.16}$$

(Problem 62). The *negentropy* $h(f_{\mathrm{gauss}}) - h(f)$ is thus maximized by non-Gaussian f, i.e., one could set $G(s) = -\log f(s)$ and maximize it. In practice estimating f is hard.

`FastICA` instead uses computational surrogates such as $G(s) = \log\cosh(s)$ or $G(s) = -e^{-s^2}$, solving $G'(s) = g(s) = 0$ via Newton–Raphson; see [9] for details.

As a sanity check, we run ICA with `FastICA` and visualize the steps.

```
import matplotlib.pyplot as plt
from sklearn.decomposition import FastICA
import numpy as np

rng = np.random.default_rng(42)

# Independent non-Gaussian sources (e.g., uniform)
S = rng.random((5000, 2))  # 5000 x 2

# Mixing matrix
A = np.array([[1, 1],
              [-1, 3]])

# Observed mixtures
X = S @ A  # 5000 x 2

# ICA (sklearn performs centering + whitening internally)
ica = FastICA(
    n_components=2,
    fun='logcosh',
    algorithm='parallel',
    max_iter=200,
    tol=1e-4,
    random_state=42
)

S_hat = ica.fit_transform(X)  # estimated independent components (5000 x 2)
# mixing_ is the linear part of the inverse transform (X = S_hat @ mixing_.T
   + mean_)
A_hat = ica.mixing_

# If you want to visualize post-centering and whitening yourself:
X_centered = X - ica.mean_
# Symmetric PCA whitening (demo)
cov = (X_centered.T @ X_centered) / X_centered.shape[0]
evals, evecs = np.linalg.eigh(cov)
W = evecs @ np.diag(1.0 / np.sqrt(evals)) @ evecs.T
X_white = X_centered @ W

fig, axes = plt.subplots(1, 3, figsize=(12, 4))
axes[0].scatter(X_centered[:, 0], X_centered[:, 1], s=5)
axes[0].set_title("After centering")

axes[1].scatter(X_white[:, 0], X_white[:, 1], s=5)
axes[1].set_title("After whitening (PCA space)")

axes[2].scatter(S_hat[:, 0], S_hat[:, 1], s=5)
axes[2].set_title("Estimated independent components (ICA)")

plt.tight_layout()
plt.show()
```

Figure 6.4 shows the output.

We can also separate synthetic mixed signals back into the original independent sources (Problem 63): Generate a sine wave and a linearly changing signal, mix them via a mixing matrix to form observed mixtures, apply FastICA to estimate independent components, and visualize the originals, mixtures, and separated signals (Fig. 6.5).

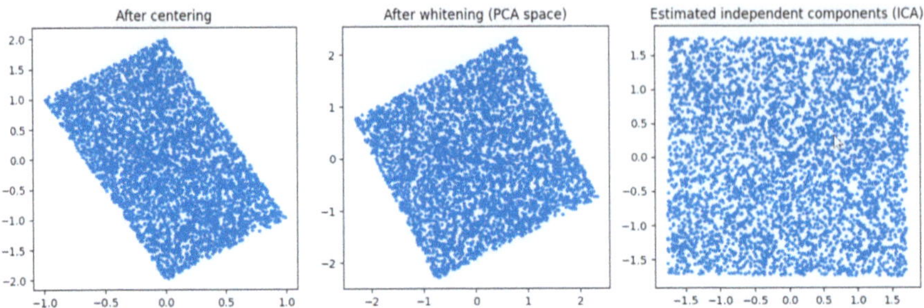

Fig. 6.4 Running ICA and visualizing the pipeline: centering/normalization, whitening, and ICA

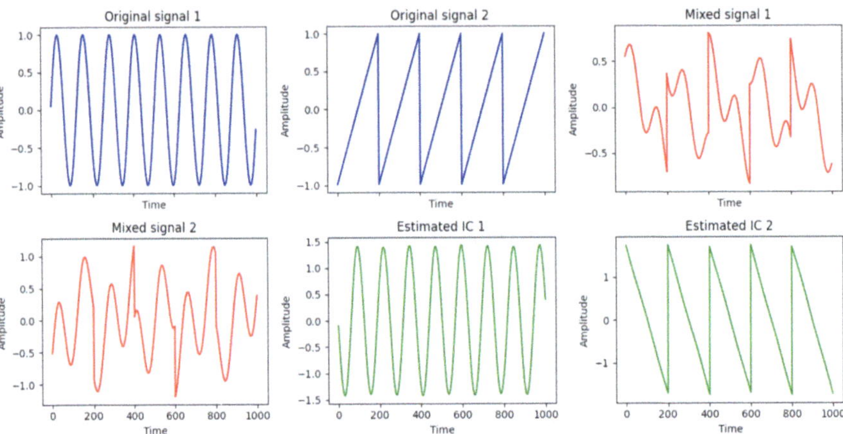

Fig. 6.5 Example of separating synthetic waveforms

6.5 ICA LiNGAM

ICA LiNGAM uses ICA to extract independent components and then determines a causal order.

Let `FastICA` yield weights `W` on the whitened data $\tilde{X} = XK$ with $K = UD^{-1/2}$:

$$S = \tilde{X}W = (XK)W = X \cdot (KW).$$

For brevity write $W := KW$ below.

In ICA the order of extracted components is arbitrary, but per (6.14) we expect $W = I - B$ (with B strictly lower triangular). The vector e is the vector of independent components. Thus we seek a row permutation of W that makes it lower triangular with ones on the diagonal.

A permutation matrix P (with exactly one 1 per row/column) permutes the rows when multiplied on the left. Determining P is equivalent to determining a bijection $\pi : \{1, \ldots, p\} \to \{1, \ldots, p\}$, of which there are $p!$. Hence choosing P is equivalent to choosing a row order (Problem 65).

Multiplying W by P on the left permutes its rows. Under a correct estimate, the diagonal entries of W are nonzero. ICA LiNGAM chooses P to minimize the sum of the reciprocals of the absolute diagonal entries of PW, i.e., minimizes $\sum_i |(PW)_{ii}|^{-1}$.

Next, let $PW = (c_{ij})$ and scale rows so that the diagonal becomes 1. With $D = \mathrm{diag}(c_{11}, \ldots, c_{pp})$, set

$$D^{-1}PW = \begin{bmatrix} 1 & \frac{c_{12}}{c_{11}} & \cdots & \frac{c_{1p}}{c_{11}} \\ \frac{c_{21}}{c_{22}} & 1 & \cdots & \frac{c_{2p}}{c_{22}} \\ \vdots & \vdots & \ddots & \vdots \\ \frac{c_{p1}}{c_{pp}} & \frac{c_{p2}}{c_{pp}} & \cdots & 1 \end{bmatrix}.$$

Assuming the above-diagonal entries are near zero, subtract the identity to obtain an estimate of B via $\widehat{I - B}$.

Because there are $p!$ possible permutations, directly minimizing over all P is infeasible for large p. Instead, reduce to an assignment problem $\min_\pi \sum_{i=1}^{p} |W_{i,\pi(i)}|^{-1}$, and solve with `SciPy`'s `linear_sum_assignment` (Hungarian method).

Finally, we can also run the `lingam` package from PyPI.

Example 6.5 (Causal Discovery with LiNGAM (Two Variables)) We generate two variables X_1, X_2 with ground truth $X_2 \to X_1$, apply LiNGAM, and check the agreement between the estimated and true DAG.

```python
import numpy as np
from lingam import DirectLiNGAM  # pip install lingam

# ---- Data generation (example: x2 → x1) ----
np.random.seed(1234)
n = 500
eps1 = np.sign(np.random.randn(n)) * np.sqrt(np.abs(np.random.randn(n)))
eps2 = np.random.rand(n) - 0.5

x2 = 3 + eps2
x1 = 0.9 * x2 + 7 + eps1
X = np.column_stack([x1, x2])

# True causal structure (column j → row i)
true_DAG = np.array([[0, 1],
                     [0, 0]])
print("True DAG:\n", true_DAG)

# ---- LiNGAM ----
model = DirectLiNGAM()
model.fit(X)
B_hat = model.adjacency_matrix_

# Estimated causal structure (column j → row i)
est_DAG = (np.abs(B_hat) > 1e-6).astype(int)
print("Estimated DAG:\n", est_DAG)
```

Please run the program yourself (Problem 64).

The estimated causal structure matches the true DAG. The estimated bias terms and noise scales are also close to the true values, illustrating that under non-Gaussianity LiNGAM can recover the correct causal structure. ■

Example 6.6 (Causal Discovery with LiNGAM (Four Variables)) We consider four variables: x_2 is upstream, x_1 and x_3 are influenced by x_2, and x_4 is influenced by x_1 and x_3:

$$x_2 \to x_1, \quad x_2 \to x_3, \quad x_1 \to x_4, \quad x_3 \to x_4.$$

We apply LiNGAM and compare the estimated structure to the true one.

```
np.random.seed(123)
n = 500

eps1 = np.sign(np.random.randn(n)) * np.sqrt(np.abs(np.random.randn(n)))
eps2 = np.random.rand(n) - 0.5
eps3 = np.sign(np.random.randn(n)) * np.abs(np.random.randn(n))**(1/3)
eps4 = (np.random.randn(n)**2 - np.mean(np.random.randn(n)**2))

x2 = eps2
x1 = 0.9 * x2 + eps1
x3 = 0.8 * x2 + eps3
x4 = -x1 - 0.9 * x3 + eps4

X = np.column_stack([x1, x2, x3, x4])

true_DAG = np.array([
    [0, 1, 0, 0],  # x2 → x1
    [0, 0, 0, 0],
    [0, 1, 0, 0],  # x2 → x3
    [1, 0, 1, 0],  # x1 → x4, x3 → x4
], dtype=int)
print("True DAG:\n", true_DAG)

model = DirectLiNGAM()
model.fit(X)
B_hat = model.adjacency_matrix_
est_DAG = (np.abs(B_hat) > 1e-6).astype(int)
print("Estimated DAG:\n", est_DAG)
```

Please run the program yourself (Problem 64).

The estimated structure largely matches the true DAG: x_4 is correctly identified as influenced by x_1 and x_3, demonstrating LiNGAM's effectiveness under non-Gaussianity. ■

6.6 Handling Confounding via a Shortest-Path Approach

LiNGAM assumes that in (6.1), (6.2), at least one of $e_1 \perp\!\!\!\perp e_2$ or $e'_1 \perp\!\!\!\perp e'_2$ holds. If neither holds (i.e., there is confounding), how can we estimate a causal order?

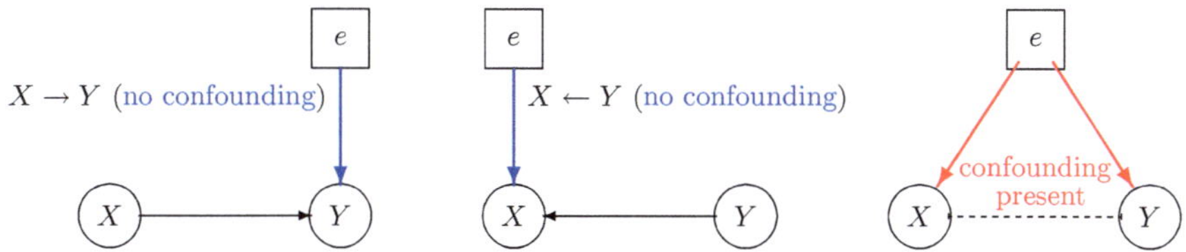

Fig. 6.6 When noise affects only a single variable (shown in blue), LiNGAM can identify the causal direction as either $X \to Y$ or $Y \to X$, provided that at least one of X or Y is non-Gaussian. When noise affects multiple variables (shown in red), the causal direction cannot be identified

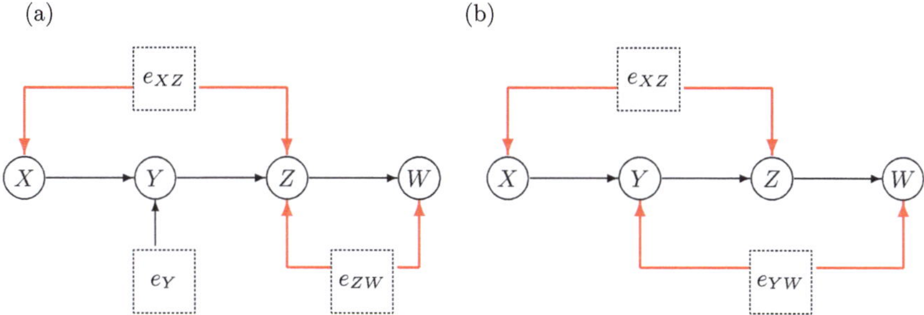

Fig. 6.7 In (**a**), $x^n \perp\!\!\!\perp y^n_x$ and $y^n_x \perp\!\!\!\perp z^n_{xy}$ hold, but $x^n \perp\!\!\!\perp z^n_{xy}$ does not. Therefore, only $X \to Y \to Z$ can be identified. In (**b**), since $x^n \perp\!\!\!\perp y^n_x$, $y^n_x \perp\!\!\!\perp z^n_{xy}$, and $z^n_{xy} \perp\!\!\!\perp w^n_{xyz}$ all hold, $X \to Y \to Z \to W$ can be identified

Here, **confounding** refers to situations where a noise source affects multiple variables and thus obscures directionality; such a source is a **confounding factor** (Fig. 6.6).

Example 6.7 In Fig. 6.7a, computing x^n, y^n_x, z^n_{xy}, the confounder e_{XZ} prevents $x^n \perp\!\!\!\perp z^n_{xy}$. However, if $x^n \perp\!\!\!\perp y^n_x$ and $y^n_x \perp\!\!\!\perp z^n_{xy}$ hold, then $X \to Y \to Z$ is identifiable. Still, due to confounder e_{ZW}, $z^n_{xy} \perp\!\!\!\perp w^n_{xyz}$ fails. In contrast, in Fig. 6.7b, if

$$x^n \perp\!\!\!\perp y^n_x, \quad y^n_x \perp\!\!\!\perp z^n_{xy}, \quad z^n_{xy} \perp\!\!\!\perp w^n_{xyz},$$

then $X \to Y \to Z \to W$ is identifiable. ■

Example 6.7 shows that testing independence only between residuals of adjacent variables suffices to identify an order. Even with confounding, as long as it does not occur between adjacent variables, the standard LiNGAM procedure applies.

Proposition 6.4 (Wang–Drton, 2023 [38]) *Even if there is a confounding factor for two variables, as long as it does not confound adjacent variables, the usual LiNGAM procedure applies.*

In practice, however, confounding often occurs between adjacent variables, so Proposition 6.4 is rarely helpful. For $p = 2$, practitioners often compare $X \to Y$ vs. $Y \to X$ using HSIC or mutual information.

For general p, how should we proceed?

Recall ICA: FastICA seeks to maximize total negentropy; Hyvärinen [9] argues that, when extracting multiple components, one should maximize the sum

$$\text{const} - \sum_{i=1}^{p} h(f_i).$$

In our notation, let $f_{1,\ldots,p}(e_1, \ldots, e_p)$ be the joint density of residuals $e_1, \ldots, e_p$ under a given order of $X_1, \ldots, X_p$. Define

$$KL(e_1, \ldots, e_p) := \mathbb{E}\left[\log \frac{f_{1,\ldots,p}(e_1, \ldots, e_p)}{f_1(e_1)\cdots f_p(e_p)}\right] \tag{6.17}$$

$$= \int \cdots \int f_{1,\ldots,p}(e_1, \ldots, e_p) \log \frac{f_{1,\ldots,p}(e_1, \ldots, e_p)}{f_1(e_1)\cdots f_p(e_p)} de_1 \cdots de_p,$$

which depends on the variable order. In ICA, (6.17) is the *multi-information* (mutual information for p variables): the KL divergence between the joint $f_{1,\ldots,p}$ and the product of marginals $f_1 \cdots f_p$.

From the above, "no confounding" is equivalent to (6.17) being zero. LiNGAM can search orders regardless of confounding, but different orders yield different (6.17) values.

We thus define the degree of confounding by (6.17). If there is no confounding, then $P(e_1, \ldots, e_p) = P(e_1) \cdots P(e_p)$ and $KL(e_1, \ldots, e_p) = 0$—exactly the LiNGAM assumption for a given order.

Recall $I(X, Y) \geq 0$ and

$$I(X, Y) = 0 \iff X \perp\!\!\!\perp Y.$$

Hence, confounding can be expressed as a telescoping sum of mutual information:

$$KL(e_1, \ldots, e_p) = \mathbb{E}\left[\log \frac{P(e_1, \ldots, e_p)}{P(e_1)P(e_2, \ldots, e_p)}\right] + \mathbb{E}\left[\log \frac{P(e_2, \ldots, e_p)}{P(e_2)\cdots P(e_3, \ldots, e_p)}\right] + \cdots$$
$$+ \mathbb{E}\left[\log \frac{P(e_{p-1}, e_p)}{P(e_{p-1})P(e_p)}\right]. \tag{6.18}$$

Since mutual information is nonnegative, the following are equivalent:

$$\begin{aligned}
\text{no confounding} &\iff KL(e_1, \ldots, e_p) = 0 \\
&\iff I(e_1, \{e_2, \ldots, e_p\}) = \cdots = I(e_{p-1}, e_p) = 0 \\
&\iff e_1, \ldots, e_p \text{ are mutually independent.}
\end{aligned}$$

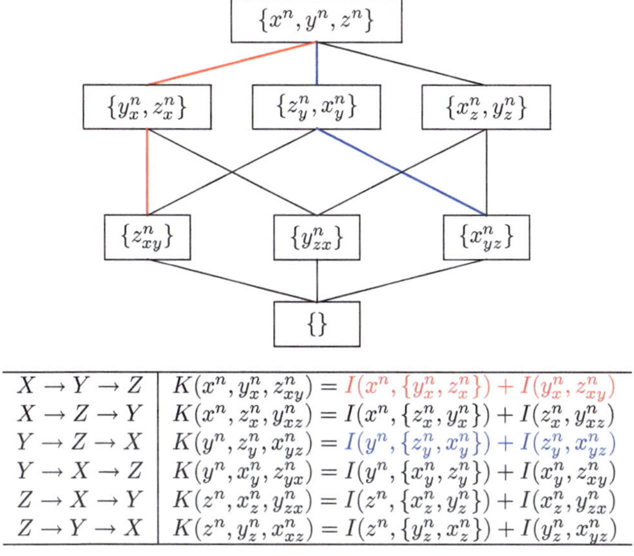

$X \to Y \to Z$	$K(x^n, y^n_x, z^n_{xy}) = I(x^n, \{y^n_x, z^n_x\}) + I(y^n_x, z^n_{xy})$
$X \to Z \to Y$	$K(x^n, z^n_x, y^n_{xz}) = I(x^n, \{z^n_x, y^n_x\}) + I(z^n_x, y^n_{xz})$
$Y \to Z \to X$	$K(y^n, z^n_y, x^n_{yz}) = I(y^n, \{z^n_y, x^n_y\}) + I(z^n_y, x^n_{yz})$
$Y \to X \to Z$	$K(y^n, x^n_y, z^n_{yx}) = I(y^n, \{x^n_y, z^n_y\}) + I(x^n_y, z^n_{xy})$
$Z \to X \to Y$	$K(z^n, x^n_z, y^n_{zx}) = I(z^n, \{x^n_z, y^n_z\}) + I(x^n_z, y^n_{zx})$
$Z \to Y \to X$	$K(z^n, y^n_z, x^n_{xz}) = I(z^n, \{y^n_z, x^n_z\}) + I(y^n_z, x^n_{yz})$

Fig. 6.8 There are six paths corresponding to the six possible orders. Along each path (order), the total distance from the top $\{x^n, y^n, z^n\}$ to the bottom $\{\}$ is compared

Wes now minimize the sum in (6.18) by constructing an ordered graph and solving a **shortest-path problem**.

For $p = 3$ variables X, Y, Z, there are 6 orders. Construct the graph in Fig. 6.8. For example, to compare $X \to Y \to Z$ vs. $Y \to Z \to X$, compute the total distances along the red and blue paths, respectively.

Given data $\{x^n, y^n, z^n\}$, compute the edge lengths as estimates of mutual information (assume $I_n(\{x^n_{yz}\}, \{\}) = I_n(\{y^n_{zx}\}, \{\}) = I_n(\{z^n_{xy}\}, \{\}) = 0$).

Treat these MI estimates as distances. For each node v, compute the best-known distance $d(v)$ from $\{x^n, y^n, z^n\}$ to v (choosing the shortest among multiple paths).

For instance, the path $\{x^n, y^n, z^n\} \to \{y^n_x, z^n_x\} \to \{z^n_{xy}\} \to \{\}$ has the total length

$$I_n(x^n, \{y^n_x, z^n_x\}) + I_n(y^n_x, z^n_{xy}) + 0 \;=\; I_n(x^n, \{y^n_x, z^n_{xy}\}) + I_n(y^n_x, z^n_{xy}),$$

which equals the estimated **KL divergence** of the noise set $\{e_1, e_2, e_3\}$ under the model $X = e_1, Y = aX + e_2, Z = bX + cY + e_3$. Our goal is the shortest path from $\{x^n, y^n, z^n\}$ to $\{\}$.

As in Fig. 6.8, first compute edge lengths from $\{X, Y, Z\}$ to $\{Y, Z\}, \{Z, X\}, \{X, Y\}$:

$$d(\{Y, Z\}) := I_n(x^n, \{y^n_x, z^n_x\}),\; d(\{Z, X\}) := I_n(y^n, \{z^n_y, x^n_y\}),\; d(\{X, Y\}) := I_n(z^n, \{x^n_z, y^n_z\}).$$

Move $\{X, Y, Z\}$ to CLOSE, and add $\{Y, Z\}, \{Z, X\}, \{X, Y\}$ to OPEN (Fig. 6.9a).

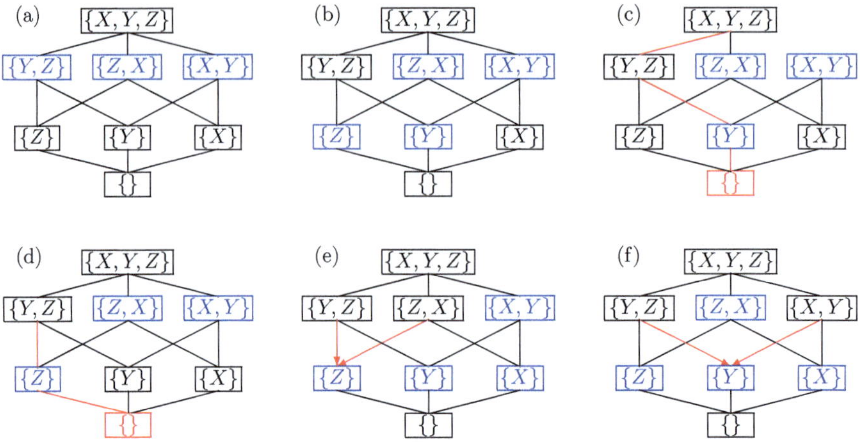

Fig. 6.9 Searching causal orders via a shortest-path problem

If $d(\{Y, Z\})$ is minimal, compute $I_n(y^n_x, z^n_{xy})$ and $I_n(z^n_x, y^n_{zx})$, and update

$$d(\{Z\}) := d(\{Y, Z\}) + I_n(y^n_x, z^n_{xy}), \qquad d(\{Y\}) := d(\{Y, Z\}) + I_n(z^n_x, y^n_{zx}). \tag{6.19}$$

Move $\{Y, Z\}$ to CLOSE, and add $\{Z\}, \{Y\}$ to OPEN (Fig. 6.9b).

Next, among $d(\{Y\}), d(\{Z\}), d(\{Z, X\}), d(\{X, Y\})$, if $d(\{Y\})$ is minimal, the shortest path is Fig. 6.9c, yielding the order $X \to Z \to Y$. If $d(\{Z\})$ is minimal, the shortest path is Fig. 6.9d, yielding $X \to Y \to Z$.

If in Fig. 6.9b $d(\{Z, X\})$ is minimal, compute $I_n(z^n_y, x^n_{yz})$ and $I_n(x^n_y, z^n_{xy})$, and update

$$d(\{X\}) := d(\{Z, X\}) + I_n(z^n_y, x^n_{yz}), \qquad d(\{Z\}) := d(\{Z, X\}) + I_n(x^n_y, z^n_{xy}). \tag{6.20}$$

Then CLOSE $\{Z, X\}$ and OPEN $\{X\}, \{Z\}$. Because (6.19) and (6.20) may compete, if the value from (6.20) is smaller, replace the previous $d(\{Z\})$ (Fig. 6.9e).

Finally, if in Fig. 6.9b $d(\{X, Y\})$ is minimal, proceed to Fig. 6.9f; competing path lengths to $\{Y\}$ are resolved by choosing the shorter path from $\{X, Y, Z\}$.

Repeat until $d(\{\})$ is obtained and the shortest path from $\{X, Y, Z\}$ to $\{\}$ is found. We call this procedure LiNGAM-MMI. It applies to any $p \geq 3$.

[Determining a causal order via shortest-path search]

1. Initialization:

   - $OPEN := \{TOP\}$, $CLOSE := \{\}$, $path(TOP) := ()$, $r(TOP) := DATA$, $d(TOP) = 0$.

2. Loop:
   (a) Select $v \in OPEN$ with minimal $d(v)$; move it to $CLOSE$.
   (b) For each neighbor $v_1, \ldots, v_m$ of v:
      i. If $v_i \notin OPEN$, compute residuals $r(v_i)$ from $r(v)$.
      ii. Compute mutual information mi from $r(v)$ and $r(v_i)$.
      iii. If $v_i \notin OPEN$ or ($v_i \in OPEN$ and $d(v) + mi < d(v_i)$), set $d(v_i) = d(v) + mi$ and $path(v_i) = append(path(v), v_i)$.
      iv. If $v_i \notin OPEN$, insert v_i into $OPEN$.
   (c) If $BOTTOM \in OPEN$, output $SHORTEST\_PATH = append(path(v), \{\})$ and stop.

Here TOP and $BOTTOM$ are the top and bottom nodes, and for nodes $u_1, \ldots, u_{s+1}$ we define

$$append((u_1, \ldots, u_s), u_{s+1}) := (u_1, \ldots, u_s, u_{s+1}).$$

We have the following proposition. Wang–Drton [38] handles certain confounding but requires time exponential in p even without confounding. Since confounding is unknown from data alone, Proposition 6.5 guarantees polynomial time in the confounding-free case.

Proposition 6.5 (Suzuki–Yang 2024 [36]) *In the shortest-path framework, if there is no confounding between adjacent variables and the LiNGAM solution is unique, then with probability one the causal order is identified after*

$$p + (p-1) + \cdots + 1 = \frac{p(p+1)}{2} \tag{6.21}$$

mutual-information computations.

Proof At the TOP, compute p MI values; the minimum is 0. For the newly OPENed $p-1$ nodes, compute MI again; a zero appears. Repeat until BOTTOM. Both the number of MI computations and comparisons total $p(p+1)/2$. ■

Minimizing (6.18) selects, among the $p!$ confounding-free statistical models, the one with minimal Kullback–Leibler divergence.

Even without confounding, this has benefits. With only data in hand, "identification" is really estimation; greedy top-down ordering via HSIC may pick a wrong variable. The shortest-path approach searches orders globally and returns a globally optimal order.

For numerical experiments and further details, see [36].

Appendix

Proof of (6.9)

First,

$$\left(y^n - \frac{c(x^n, y^n)}{v(x^n)}x^n\right)_z = \frac{c\left(y^n - \frac{c(x^n, y^n)}{v(x^n)}x^n, z^n\right)}{v\left(y^n - \frac{c(x^n, y^n)}{v(x^n)}x^n\right)}\left\{y^n - \frac{c(x^n, y^n)}{v(x^n)}x^n\right\},$$

$$\begin{aligned} c\left(y^n - \frac{c(x^n, y^n)}{v(x^n)}x^n, z^n\right) &= c(y^n, z^n) - \frac{c(x^n, y^n)c(x^n, z^n)}{v(x^n)} \\ &= \frac{v(x^n)c(y^n, z^n) - c(x^n, y^n)c(x^n, z^n)}{v(x^n)}, \\ v\left(y^n - \frac{c(x^n, y^n)}{v(x^n)}x^n\right) &= v(y^n) - 2\frac{c(x^n, y^n)^2}{v(x^n)} + \frac{c(x^n, y^n)^2}{v(x^n)} \\ &= \frac{v(x^n)v(y^n) - c(x^n, y^n)^2}{v(x^n)}. \end{aligned}$$

Thus z^n_{xy} becomes

$$\begin{aligned} z^n_{xy} =& z^n_x - (y^n_x)_z = (z^n - \frac{c(x^n, z^n)}{v(x^n)}x^n) - (y^n - \frac{c(x^n, y^n)}{v(x^n)}x^n)_z \\ =& z^n - \frac{c(x^n, z^n)}{v(x^n)}x^n - \frac{v(x^n)c(y^n, z^n) - c(x^n, y^n)c(x^n, z^n)}{v(x^n)v(y^n) - c(x^n, y^n)^2}\left\{y^n - \frac{c(x^n, y^n)}{v(x^n)}x^n\right\} \\ =& z^n - \frac{v(y^n)c(x^n, z^n) - c(x^n, y^n)c(y^n, z^n)}{v(x^n)v(y^n) - c(x^n, y^n)^2}x^n \\ & - \frac{v(x^n)c(y^n, z^n) - c(x^n, y^n)c(x^n, z^n)}{v(x^n)v(y^n) - c(x^n, y^n)^2}y^n, \end{aligned} \tag{6.22}$$

which is (6.9). ■

Proof of Proposition 6.1

Let

$$Y_1 = \sum_{i=1}^{m} \alpha_i X_i, \qquad Y_2 = \sum_{i=1}^{m} \beta_i X_i$$

be independent. Their joint characteristic function satisfies (by (2.26)) $\Phi_{Y_1 Y_2}(u, v) = \Phi_{Y_1}(u)\Phi_{Y_2}(v)$. Independence of $X_1, \ldots, X_m$ yields

$$\Phi_{Y_1 Y_2}(u, v) = \mathbb{E}\left[\exp\left\{i\left(u\sum_{j=1}^{m}\alpha_j X_j + v\sum_{j=1}^{m}\beta_j X_j\right)\right\}\right]$$
$$= \prod_{j=1}^{m}\mathbb{E}\big[\exp\{i(\alpha_j u + \beta_j v)X_j\}\big] = \prod_{j=1}^{m}\Phi_{X_j}(\alpha_j u + \beta_j v).$$

Hence identically,

$$\psi_{X_1}(\alpha_1 u + \beta_1 v) + \cdots + \psi_{X_m}(\alpha_m u + \beta_m v) = \psi_{Y_1}(u) + \psi_{Y_2}(v), \tag{6.23}$$

where $\psi_{X_i} = \log \Phi_{X_i}$ and $\psi_{Y_j} = \log \Phi_{Y_j}$.[2]

Move terms with $\alpha_j = 0$ or $\beta_j = 0$ to the RHS and write

$$\sum_{j=1}^{n}\Delta_j^{(0)}(u, v) = A_0(u) + B_0(v), \tag{6.24}$$

with $n \le m$. If all $\alpha_j\beta_j \neq 0$, then $\Delta_j^{(0)}(u, v) = \psi_{X_j}(\alpha_j u + \beta_j v)$, $A_0 = \psi_{Y_1}$, $B_0 = \psi_{Y_2}$ $(n = m)$. Fix (r, s) with $\alpha_1 r + \beta_1 s = 0$ and replace (u, v) by $(u+r, v+s)$; subtract the original (6.24):

$$\sum_{j=2}^{n}\Delta_j^{(1)}(u, v) = A_1(u) + B_1(v),$$

with suitable definitions of $\Delta_j^{(1)}$, A_1, B_1. Repeating for $j = 2, \ldots, n-1$ yields

$$\Delta_n^{(n-1)}(u, v) = A_{n-1}(u) + B_{n-1}(v).$$

Differencing once in u and once in v gives $\Delta_n^{(n+1)}(u, v) = 0$, so $\Delta_j^{(0)}$ vanishes after at most $n + 1$ derivatives and is thus a polynomial of degree $\le n$. Therefore, for each j with $\alpha_j\beta_j \neq 0$, ψ_{X_j} is a polynomial of degree $\le n$.

We use the following lemma (see [15], §3.3 "Marcinkiewicz's Theorem"; also [11], §2.5):

[2] Characteristic functions satisfy $\Phi(0) = 1$ and are continuous; hence $\log \Phi(t)$ is finite in a neighborhood of 0.

Lemma 6.1 (Marcinkiewicz, 1938 [16]) *If a characteristic function can be written as* $\Psi(t) = e^{P(t)}$ *with polynomial* P*, then* $\deg(P) \le 2$.

If $\deg(P) = 2$ with positive leading coefficient, $\log \Psi$ would be unbounded; also $\Psi(0) = 1$ implies the constant term is 0. Hence, for any k with $\alpha_k \beta_k \neq 0$, we must have $\psi_{X_k}(t) = i\mu t - \sigma^2 t^2/2$ with $\sigma^2 > 0$ (else X_k is degenerate), i.e., Φ_{X_k} is Gaussian and X_k is normal. ■

Proof of Proposition 6.3

As an illustration, suppose the true order is $X_1 \to X_2 \to X_3 \to X_4 \to X_5 \to X_6$ and the joint density factorizes as

$$f(x_1, x_2, x_3, x_4, x_5, x_6) = f(x_1, x_2) \frac{f(x_3, x_4, x_5) f(x_3, x_4, x_6)}{f(x_3, x_4)}.$$

Then orders such as

$$X_3 \to X_4 \to X_5 \to X_6 \to X_1 \to X_2, \quad X_1 \to X_2 \to X_3 \to X_4 \to X_6 \to X_5$$

are also valid. We consider independent variable groups (e.g., $\{X_1, X_2\}$ vs. $\{X_3, X_4, X_5, X_6\}$) and parallel subgroups (e.g., $\{X_3, X_4, X_5\}$ vs. $\{X_3, X_4, X_6\}$); if each is identifiable, we regard the whole as identifiable. Thus we may assume without loss that all SEM coefficients

$$X_i = \sum_{j=1}^{i-1} \beta_{i,j} X_j + e_i$$

are nonzero ($j = 1, \ldots, i-1, i = 1, \ldots, p$).

For each $i = 2, \ldots, p$, by $e_1 \perp\!\!\!\perp X_j$ ($j = 2, \ldots, i-1$) and $e_1 \perp\!\!\!\perp e_i$, we have that $X_1 = e_1$ is independent of the residual

$$X_i - \beta_{i,1} X_1 = \sum_{j=2}^{i-1} \beta_{i,j} X_j + e_i.$$

Conversely, suppose there exists $\alpha \neq 0$ with $X_i \perp\!\!\!\perp (e_1 - \alpha X_i)$—i.e., some upstream X_i (not X_1) is independent of the residual $X_1 - \alpha X_i$ (if $\alpha = 0$, then $X_i \perp\!\!\!\perp X_1$, contradicting our factorization assumption). For $i \geq 3$ (the $i = 2$ case is already handled), there exist constants $\gamma_1, \ldots, \gamma_i$ such that

$$X_i = \sum_{j=1}^{i-2} \gamma_j e_j + \beta_{i,i-1} e_{i-1} + e_i,$$

$$e_1 - \alpha X_i = (1 - \alpha\gamma_1) e_1 - \alpha \sum_{j=2}^{i-2} \gamma_j e_j - \alpha\beta_{i,i-1} e_{i-1} - \alpha e_i.$$

(For example, when $i = 3$, $X_3 = (\beta_{3,1} + \beta_{3,2}\beta_{2,1})e_1 + \beta_{3,2}e_2 + e_3$, so $\gamma_1 = \beta_{3,1} + \beta_{3,2}\beta_{2,1}$ and $\gamma_2 = \beta_{3,2}$.) Since α, $\beta_{i,i-1} \neq 0$, applying Proposition 6.1 forces e_{i-1}, e_i to be Gaussian, contradicting the non-Gaussianity assumption. ∎

Problems 54–69

54. Why does the fact that the value of (6.9) does not change when swapping x^n and y^n imply $z_{xy}^n = z_{yx}^n$?
55. Fill in the blanks below to complete `search_pval`, a function that finds a variable ordering based on p-values.

```
def search_pval(Z, index=None, sigma2=1.0, kernel=None, alpha=0.05):
    m = len(Z)
    if index is None:
        index = list(range(m))
    if m < 2:
        return index[:]

    import math
    best_p = -math.inf
    best_W = None
    best_k = None

    for i in range(m):
        # Blank (1)
        res = LiNGAM_2(i, Z, proc="p.val", sigma2=sigma2, kernel=kernel,
            alpha=alpha)

        # Blank (2)
        if res["p.val"] > best_p:
            # Blank (3)
            best_p = res["p.val"]
            best_W = res["Z"]
            best_k = index[i]

    # recursion
    next_index = [j for j in index if j != best_k]
    # Blank (4)
    tail = search_pval(best_W, next_index, sigma2=sigma2, kernel=kernel,
        alpha=alpha)
    return [best_k] + tail
```

56. As in Example 6.3, apply `search_HSIC` to the dataset `crime.txt` https://hastie.su.domains/StatLearnSparsity_files/DATA/crime.html, and determine a causal ordering among the seven variables. The meanings of the variables are:

X1	Total crime rate per million population
X2	Violent crime rate per 100,000 population
X3	Annual police budget per capita
X4	% of persons age 25+ with 4 years of high school
X5	% of persons age 16–19 not in high school and not high school graduates
X6	% of persons age 18–24 in college
X7	% of persons age 25+ who are college graduates

57. In the discussion right after Proposition 6.1, show that

$$a = 0 \Longleftrightarrow a' = 0 \Longleftrightarrow X \perp\!\!\!\perp Y.$$

Also, explain why $aa' = 1$ cannot occur as long as $\sigma_2^2 > 0$.

58. In (6.10), under the assumptions $cov(e_1, e_2) = cov(e_1', e_2') = 0$, show that if $e_1 \perp\!\!\!\perp e_2$ and e_1, e_2 are Gaussian, then $e_1' \perp\!\!\!\perp e_2'$ and e_1', e_2' are also Gaussian.

59. Consider identifiability when the true order $X \to Y \to Z$ (residuals: e_1, e_2, e_3) is viewed as an alternative order $Y \to Z \to X$ (residuals: e_1', e_2', e_3').

   a. Show that when

$$\begin{cases} X = e_1 \\ Y = aX + e_2 \\ Z = bX + cY + e_3 \end{cases}, \qquad \begin{cases} Y = e_1' \\ Z = a'Y + e_2' \\ X = b'Y + c'Z + e_3', \end{cases}$$

   the residuals satisfy

$$\begin{aligned} e_1' &= ae_1 + e_2, \\ e_2' &= \{b + a(c - a')\}e_1 + (c - a')e_2 + e_3, \\ e_3' &= \{1 - bc' - a(b' + cc')\}e_1 - (b' + cc')e_2 - c'e_3. \end{aligned}$$

   b. Assuming zero pairwise correlations among e_1', e_2', e_3', eliminate a', b', c' in favor of a, b, c, and show that

$$\begin{aligned} e_1' &= ae_1 + e_2, \\ e_2' &= \frac{b\sigma_2^2}{a^2\sigma_1^2 + \sigma_2^2}e_1 - \frac{ab\sigma_1^2}{a^2\sigma_1^2 + \sigma_2^2}e_2 + e_3, \\ e_3' &= \frac{\sigma_2^2\sigma_1^2}{b^2\sigma_1^2\sigma_2^2 + (a^2\sigma_1^2 + \sigma_2^2)\sigma_3^2}e_1 - \frac{a\sigma_1^2\sigma_3^2}{b^2\sigma_1^2\sigma_2^2 + (a^2\sigma_1^2 + \sigma_2^2)\sigma_3^2}e_2 \\ &\quad - \frac{b\sigma_1^2\sigma_2^2}{b^2\sigma_1^2\sigma_2^2 + (a^2\sigma_1^2 + \sigma_2^2)\sigma_3^2}e_3. \end{aligned}$$

c. If $a, b, c \neq 0$ and at least one of e_1, e_2, e_3 is non-Gaussian, show that **no pair** among e_1', e_2', e_3' is independent.
d. Show that if $Y = ae_1 + e_2$ is non-Gaussian, then at least one of e_1, e_2 is non-Gaussian, and if $Z = be_1 + c(ae_1 + e_2) + e_3$ is non-Gaussian, then at least one of e_1, e_2, e_3 is non-Gaussian.
e. Conclude that if at least one of X, Y, Z is non-Gaussian, then e_1', e_2', e_3' cannot be mutually independent; that is, the true order $X \to Y \to Z$ will not be misidentified as $Y \to Z \to X$.

Hint: True model: $X = e_1,\ Y = aX + e_2,\ Z = bX + cY + e_3$. Wrong model: $Y = e_1',\ Z = a'Y + e_2',\ X = b'Y + c'Z + e_3'$. We have $e_1' = Y = ae_1 + e_2$, $e_2' = Z - a'Y$, $e_3' = X - b'Y - c'Z$. Choose a', b', c' so that covariances are zero. Focus on independence of residuals, and use that identifiability fails without non-Gaussianity.

60. Consider the true order $X \to Y \to Z$ (residuals: e_1, e_2, e_3) and two wrong orders $Y \to X \to Z$ (residuals: e_1', e_2', e_3') and $X \to Z \to Y$ (residuals: e_1'', e_2'', e_3'').

a. Show that they can be written as

$$\begin{cases} Y = e_1' \\ X = a'Y + e_2' \\ Z = b'Y + c'X + e_3' \end{cases}, \qquad \begin{cases} X = e_1'' \\ Z = a''X + e_2'' \\ Y = b''X + c''Z + e_3'' \end{cases}$$

b. By requiring zero pairwise covariances among $\{e_1', e_2', e_3'\}$ and eliminating a', b', c', and similarly eliminating a'', b'', c'' for $\{e_1'', e_2'', e_3''\}$, show

$$\begin{cases} e_1' = ae_1 + e_2 \\ e_2' = \dfrac{\sigma_2^2}{a^2\sigma_1^2 + \sigma_2^2} e_1 - \dfrac{a\sigma_1^2}{a^2\sigma_1^2 + \sigma_2^2} e_2 \\ e_3' = e_3 \end{cases}, \qquad \begin{cases} e_1'' = e_1 \\ e_2'' = ce_2 + e_3 \\ e_3'' = \dfrac{\sigma_3^2}{c^2\sigma_2^2 + \sigma_3^2} e_2 - \dfrac{c\sigma_2^2}{c^2\sigma_2^2 + \sigma_3^2} e_3 \end{cases}$$

c. Show that if e_3 is non-Gaussian and e_1, e_2 are Gaussian, misidentification as $Y \to X \to Z$ can occur.
d. Show that if e_1 is non-Gaussian and e_2, e_3 are Gaussian, misidentification as $X \to Z \to Y$ can occur.

Hint: True order: $X = e_1,\ Y = aX + e_2,\ Z = cY + e_3$. Wrong order 1: $Y = e_1',\ X = a'Y + e_2',\ Z = b'Y + c'X + e_3'$. Wrong order 2: $X = e_1'',\ Z = a''X + e_2'',\ Y = b''X + c''Z + e_3''$. Choose coefficients so covariances vanish and express e_1', e_2', e_3' via e_1, e_2, e_3. If e_3 is non-Gaussian, focus on residuals containing e_3, etc. By Darmois–Skitovich, if at least one source is non-Gaussian, linear-combination residuals sharing that source cannot be independent.

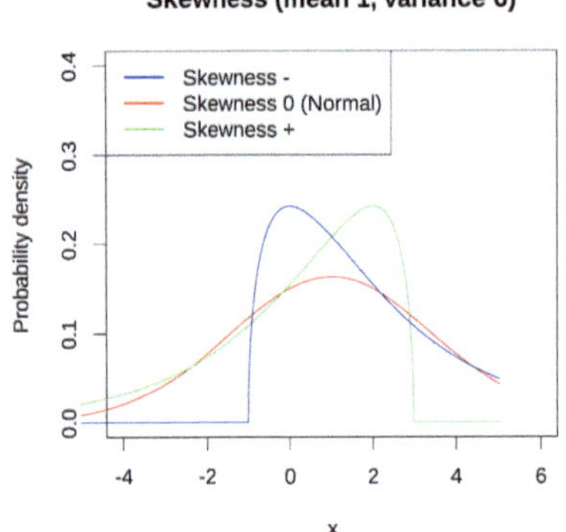

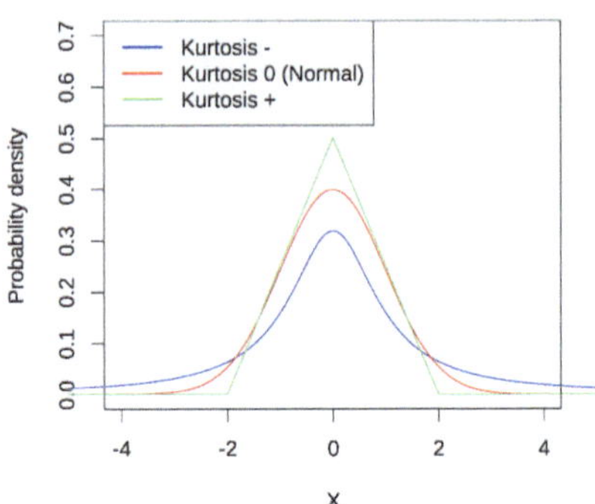

Fig. 6.10 Skewness (left) and kurtosis (right): positive, negative, and zero. For a normal distribution (red curves in both panels), both skewness and kurtosis are zero

61. Various indices measure the deviation from normality. Besides mean and variance, skewness and kurtosis are

$$\text{skewness} = \mathbb{E}_Y[(Y-\mu)^3/\sigma^3], \qquad \text{kurtosis} = \mathbb{E}_Y[(Y-\mu)^4/\sigma^4] - 3$$

(see Fig. 6.10).

a. Show that both are zero for a normal distribution.
b. In this chapter's notation, how can skewness and kurtosis be represented as choices of $G(s)$?

Hint: Skewness captures asymmetry, kurtosis sharpness. A normal $\mathcal{N}(\mu, \sigma^2)$ is symmetric; the standardized fourth moment equals 3. In ICA, non-Gaussianity measures are of the form $\mathbb{E}[G(s)]$; choose G accordingly.

62. a. Show that for $X \sim N(0, \sigma^2)$, the differential entropy equals the RHS of (6.16).
b. Show that assumption (6.15) implies inequality (6.16).
63. Reproduce Fig. 6.5 in Python by filling the blanks.

```
# pip install numpy scikit-learn matplotlib
from sklearn.decomposition import FastICA

# --- Generate independent signals S (1000×2) ---
t = np.arange(1, 1001)
s1 = np.sin(t / 20.0)                                   # signal 1:
    sine
s2 = np.tile(((np.arange(1, 201) - 100) / 100.0), 5)    # signal 2:
    repeated ramp
S = np.column_stack([s1, s2])

# --- Mixing matrix A and observed data X ---
A = np.array([[0.291, 0.6557],
              [-0.5439, 0.5572]])
X = S @ A
```

```
# --- Apply ICA ---
ica = FastICA(
    n_components=2,
    algorithm="parallel",
    fun="logcosh",
    max_iter=200,
    tol=1e-4,
    whiten="unit-variance",
    random_state=0
)
S_est = ica.fit_transform(X)  # estimated ICs (order/sign arbitrary)

# --- Visualization (2 rows × 3 cols) ---
fig, axes = plt.subplots(2, 3, figsize=(12, 6), sharex=True)

# Original signals
axes[0, 0].plot(t, BLANK(1), color="blue");  axes[0, 0].set_title("
  Original␣signal␣1")
axes[0, 1].plot(t, BLANK(2), color="blue");  axes[0, 1].set_title("
  Original␣signal␣2")

# Mixed signals
axes[0, 2].plot(t, BLANK(3), color="red");   axes[0, 2].set_title("Mixed
  ␣signal␣1")
axes[1, 0].plot(t, X[:, 1], color="red");    axes[1, 0].set_title(BLANK
  (4))

# ICA estimates
axes[1, 1].plot(t, BLANK(5), color="green"); axes[1, 1].set_title("
  Estimated␣IC␣1")
axes[1, 2].plot(t, BLANK(6), color="green"); axes[1, 2].set_title("
  Estimated␣IC␣2")

for ax in axes.ravel():
    ax.set_xlabel("Time"); ax.set_ylabel("Amplitude")

plt.tight_layout()
plt.show()
```

64. Execute the Python programs in Examples 6.5 and 6.6 and confirm the outputs.
65. For a $p \times p$ permutation matrix P:

    a. List all permutation matrices for $p = 3$, and, for each, compute PA for a generic $p \times p$ matrix $A = (a_{i,j})$.
    b. How can you permute the *columns* of A? Specify what matrix to multiply and on which side.
    c. Why are there exactly $p!$ distinct permutation matrices of size p?

    Hint: A permutation matrix has exactly one 1 in each row and column. Left-multiplying permutes rows; right-multiplying by a permutation (or left-multiplying by its transpose) permutes columns. The number of permutations of p items is $p!$.
66. In the shortest-path-based algorithm, once a vertex is moved to CLOSE, no other path will later reach it with a shorter distance. Why? *Hint:* Being added to CLOSE means it currently has the smallest tentative distance in OPEN. With nonnegative edge lengths, any other route would be no shorter; thus no update is possible afterward.
67. Prove that the third equality in (6.22) holds.

68. In the proof of Proposition 6.2 (i.e., Proposition 6.3), why is it legitimate—under that definition of identifiability—to assume $\beta_{i,j} \neq 0$? Also, how exactly is Proposition 6.1 (Darmois–Skitovich) applied?
69. On synthetic three-variable data, run (i) LiNGAM applied *after* extracting independent components with `FastICA`, and (ii) Direct LiNGAM alone. Compare the inferred orders. Through numerical experiments, discuss under what conditions (e.g., noise distribution or sample size) the two methods agree or diverge.

Chapter 7
Information Criteria and Marginal Likelihood

In this chapter, we discuss information criteria (AIC and BIC) and Bayesian marginal likelihood, which play central roles in structure learning for graphical models. We first clarify the definitions and objectives of AIC (Akaike Information Criterion) and BIC (Bayesian Information Criterion) and explain their positioning as selection methods based on the trade-off between goodness of fit and model complexity. Using linear regression models and categorical data as examples, we show how to compute these criteria concretely.

In the latter part, we introduce the definition of marginal likelihood in the Bayesian framework, provide examples using Jeffreys' prior and the inverse-Wishart prior for normal distributions, and show analytic calculations.

We then use Stirling's formula and Laplace's method to explain theoretically how BIC can be regarded as an approximation to the negative log marginal likelihood. For AIC, we clarify its connection to predictive performance.

These scores are not merely model evaluation indices; they also serve as guides for exploring the best graphical model among candidates. In the next chapter, we will leverage such scores to learn structures—namely, score-based structure learning. This chapter serves as the preparation for that.

7.1 Overview of AIC and BIC

In Chap. 8, to estimate the structure of a graphical model, we detect dependencies among random variables from observed data. To this end we use **information criteria** such as **AIC** (Akaike Information Criterion) and **BIC** (Bayesian Information Criterion).

J. Suzuki, *Graphical Models and Causal Discovery with Python*,
https://doi.org/10.1007/978-981-95-5308-2_7

When evaluating a statistical model, we must consider not only how well the data fit the model but also how simple the model is. AIC and BIC are defined by

$$AIC = \sum_{i=1}^{n}\{-\log p(x_i \mid \hat{\theta}(x_1, \ldots, x_n))\} + d \tag{7.1}$$

$$BIC = \sum_{i=1}^{n}\{-\log p(x_i \mid \hat{\theta}(x_1, \ldots, x_n))\} + \frac{d}{2}\log n, \tag{7.2}$$

where $\hat{\theta}$ is the maximum likelihood estimator (see Sect. 2.4), and d is the **number of parameters**. For instance, in linear regression, d is the number of regression coefficients (i.e., the number of explanatory variables). AIC and BIC differ only in the second term.

$$\sum_{i=1}^{n}\{-\log p(x_i \mid \hat{\theta}(x_1, \ldots, x_n))\}, \qquad d$$

represent, respectively, goodness of fit and simplicity. AIC and BIC weight the first and second terms as 1 vs. 1, and 2 vs. $\log n$, and select the model with the smaller sum.

From the definitions, it is clear that the balance between fit and simplicity differs. Just as an applicant might pass an exam at a university that weights mathematics more heavily and fail at one that weights English more, AIC tends to favor fit relative to simplicity compared with BIC and thus selects more complex models.

Example 7.1 Continuing Example 2.14, since $\theta_1, \ldots, \theta_\alpha$ sum to 1, the parameter actually moves in $(\alpha - 1)$-dimensions (Fig. 7.1). Hence,

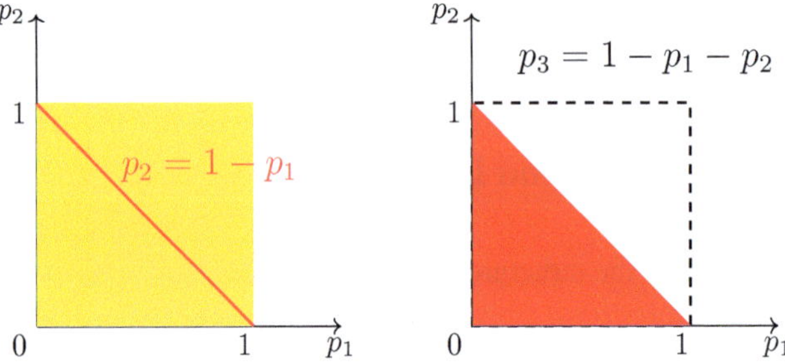

Fig. 7.1 For $\alpha = 2$, even though it appears two-dimensional, (p_1, p_2) can only move along the line segment $p_2 = 1 - p_1$ (left). For $\alpha = 3$, since $p_3 = 1 - p_1 - p_2$ (extending in the direction out of the page) is automatically determined from the values of (p_1, p_2), the interior of the red triangle can be regarded as the parameter space

$$AIC = \sum_{j=1}^{\alpha} -k_j \log \frac{k_j}{n} + \alpha - 1 \tag{7.3}$$

$$BIC = \sum_{j=1}^{\alpha} -k_j \log \frac{k_j}{n} + \frac{\alpha - 1}{2} \log n \tag{7.4}$$

■

In Python:

```
def AIC_1(k):
    k = np.asarray(k)
    n = np.sum(k)
    mask = k > 0
    return np.sum(k[mask] * np.log(n / k[mask])) + len(k) - 1
```

```
def BIC_1(k):
    k = np.asarray(k)
    n = np.sum(k)
    mask = k > 0
    return np.sum(k[mask] * np.log(n / k[mask])) + (len(k) - 1) / 2 * np.log(
      n)
```

Example:

```
n = 200
p = 0.25
x = np.random.binomial(1, p, size=n)
k = np.bincount(x)
print(AIC_1(k))
print(BIC_1(k))
```

Example 7.2 Continuing Example 2.15, there are three parameters $\beta_0, \beta_1, \sigma^2$, so

$$BIC = \frac{n}{2} \log \hat{\sigma}^2 + \frac{n}{2} \log(2\pi e) + \frac{3}{2} \log n.$$

In general, if there are p slopes (including the intercept), the number of parameters is $p + 1$ (Example 2.15 has $p = 2$). Then (2.33) still holds, and thus

$$BIC = \frac{n}{2} \log \hat{\sigma}^2 + \frac{n}{2} \log(2\pi e) + \frac{p+1}{2} \log n.$$

■

BIC is typically used to select variables based on relative magnitudes rather than absolute values. Therefore, we may ignore constants and compare after multiplying by a constant. In this sense, in linear regression one often writes

$$n \log \hat{\sigma}^2 + p \log n \tag{7.5}$$

and calls it "BIC." As long as whether the intercept is counted in p is handled consistently, there is no issue.

To compute $\hat{\sigma}^2$ concretely, we first obtain $\hat{\beta} \in \mathbb{R}^p$. In matrix form, for data $X \in \mathbb{R}^{n\times p}$, $y \in \mathbb{R}^n$,

$$\|y - X\beta\|^2$$

is minimized by

$$\hat{\beta} = (X^\top X)^{-1} X^\top y, \tag{7.6}$$

provided $(X^\top X)^{-1}$ exists, since differentiating with respect to β gives $-X^\top(y - X\beta) = 0$, i.e., $X^\top X\beta = X^\top y$. Here $\|\cdot\|^2$ denotes the sum of squares over the n elements. Generalizing (2.32), we have

$$\hat{\sigma}^2 = \|y - X\hat{\beta}\|^2/n.$$

Python implementation:

```
def sigma2(X, y):
    n = len(y)
    # Least squares using lstsq (numerically robust)
    beta_hat, *_ = np.linalg.lstsq(X, y, rcond=None)
    return np.sum((y - X @ beta_hat) ** 2) / n
```

```
def AIC_2(X, y):
    n = len(y)
    p = X.shape[1]
    s2 = sigma2(X, y)
    return (n / 2) * np.log(s2) + n / 2 * np.log(2 * np.pi * np.e) + p + 1
```

```
def BIC_2(X, y):
    n = len(y)
    p = X.shape[1]
    s2 = sigma2(X, y)
    return (n / 2) * np.log(s2) + n / 2 * np.log(2 * np.pi * np.e) + (p + 1)
      / 2 * np.log(n)
```

In practice, the following simplified forms are often used:

```
def AIC_3(X, y):
    return len(y) * np.log(sigma2(X, y)) + 2 * X.shape[1]
```

```
def BIC_3(X, y):
    return len(y) * np.log(sigma2(X, y)) + X.shape[1] * np.log(len(y))
```

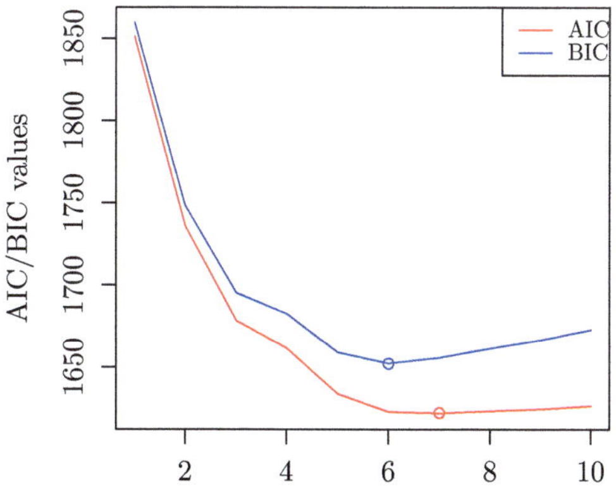

Fig. 7.2 Evaluation on the Boston dataset (Example 7.3). Although BIC's absolute values are larger, BIC favors simpler models (fewer variables)

```
# Quick check
n, p = 100, 3
X = np.random.randn(n, p)
y = np.random.randn(n)
print(AIC_2(X, y), BIC_2(X, y), AIC_3(X, y), BIC_3(X, y))
```

Note that the simplified AIC/BIC (`AIC_3`, `BIC_3`) can be negative.

Finally in this section, we empirically verify that AIC tends to select more complex models than BIC (Fig. 7.2).

Example 7.3 Using the Boston dataset (Table 5.1), we search for explanatory variables for MEDV (column 14). We run the code below: We vary the number of variables k from 1 to p and plot the minimal AIC/BIC at each k. We restrict explanatory variables to the continuous ones. We find AIC is minimized at $k = 7$, while BIC is minimized at $k = 6$.

```
import pandas as pd
import itertools

df = pd.read_csv("Boston.csv")
X = df.iloc[:, [0, 2, 4, 5, 6, 7, 9, 10, 11, 12]].to_numpy()
y = df.iloc[:, 13].to_numpy()
n, p = X.shape
X = np.column_stack([np.ones(n), X])  # add intercept

AIC_seq = []
BIC_seq = []
for k in range(1, p + 1):
    min_s = np.inf
    for comb in itertools.combinations(range(p), k):
        q = [0] + [j + 1 for j in comb]  # intercept + selected vars
        s = sigma2(X[:, q], y)
        if s < min_s:
            min_s = s
            q_min = q
    AIC_seq.append(AIC_3(X[:, q_min], y))
    BIC_seq.append(BIC_3(X[:, q_min], y))
```

```
plt.plot(range(1, p + 1), AIC_seq, color="red", label="AIC")
plt.plot(range(1, p + 1), BIC_seq, color="blue", label="BIC")
plt.xlabel("Number of variables")
plt.ylabel("AIC/BIC value")
plt.legend()
plt.show()
```

■

7.2 Marginal Likelihood (Discrete)

In Example 2.13, the second column of Table 2.1 satisfies

$$\sum_{x_1=0}^{1}\sum_{x_2=0}^{1}\sum_{x_3=0}^{1} p(x_1, x_2, x_3 \mid \theta) = 1. \tag{7.7}$$

However, since $\hat{\theta}$ depends on x_1, x_2, x_3, plugging it into the likelihood gives $p(x_1, x_2, x_3 \mid \hat{\theta}(x_1, x_2, x_3))$ (column 5 of Table 2.1), and unlike (7.7),

$$\sum_{x_1=0}^{1}\sum_{x_2=0}^{1}\sum_{x_3=0}^{1} p(x_1, x_2, x_3 \mid \hat{\theta}(x_1, x_2, x_3)) = \frac{102}{27} > 1.$$

This indicates that the MLE overreacts to the data and provides estimates inconsistent with other potential observations.

This arises because θ is unknown and we attempt to estimate it as a single value. Instead, consider setting numbers $q(x_1, x_2, x_3)$, $x_1, x_2, x_3 \in \{0, 1\}$ that satisfy

$$\sum_{x_1=0}^{1}\sum_{x_2=0}^{1}\sum_{x_3=0}^{1} q(x_1, x_2, x_3) = 1, \qquad q(x_1, x_2, x_3) \geq 0. \tag{7.8}$$

For example, using a function $\varphi(\theta)$ on θ with

$$\int_0^1 \varphi(\theta)d\theta = 1, \qquad \varphi(\theta) \geq 0,\ 0 \leq \theta \leq 1, \tag{7.9}$$

define

$$q(x_1, x_2, x_3) = \int_0^1 p(x_1, x_2, x_3 \mid \theta)\varphi(\theta)d\theta. \tag{7.10}$$

If $\varphi(\cdot)$ satisfies (7.9), then (7.8) holds. The function $\varphi(\cdot)$ is called a **prior distribution** on θ. We will use this approach to set $q(x_1, x_2, x_3)$, $x_1, x_2, x_3 \in \{0, 1\}$.

When $p(x \mid \theta)$ is defined for each parameter θ as in (7.10),

$$\int_{\Theta} \varphi(\theta) \prod_{i=1}^{n} p(x_i \mid \theta) d\theta$$

is called the **marginal likelihood** of the sequence $x_1, \ldots, x_n$ (similarly defined for continuous-valued sequences; see Sect. 7.3).

Example 7.4 Let $\varphi(\theta) = 1$ $(0 \le \theta \le 1)$. Then

$$q(0,0,0) = q(1,1,1) = \int_0^1 \theta^3 \, d\theta = \left[\frac{\theta^4}{4}\right]_0^1 = \frac{1}{4},$$

$$q(0,1,1) = q(1,0,0) = \int_0^1 (1-\theta)\theta^2 \, d\theta = \left[\frac{\theta^3}{3} - \frac{\theta^4}{4}\right]_0^1 = \frac{1}{12}.$$

By symmetry the other four cases are analogous, and (7.8) is verified. ■

Now consider the case where X takes two values. As seen in Example 7.4, one might feel it is natural to adopt a uniform prior $\varphi(\theta) = 1$—as a "noninformative" state.

However, as the next example shows, the property of being "uniform" depends on the parameterization and is only apparent.

Example 7.5 Assume θ is uniformly distributed on $[0, 1]$. Under the transformation $t = \sqrt{\theta}$, the prior for t is

$$\frac{d\theta}{dt} = 2t, \quad \varphi(\theta)d\theta = 1 \cdot d\theta = 2t \, dt,$$

i.e., $\varphi(t) = 2t$, which is not uniform. ■

Thus, uniformity is not invariant to reparameterization and cannot preserve "noninformativeness" under coordinate changes. It is natural to define a noninformative prior that is invariant to parameterization.

From this viewpoint, we introduce the invariant prior proposed by Jeffreys. For $\theta \in \Theta \subseteq \mathbb{R}^p$, the **Fisher information matrix** is

$$I_{i,j}(\theta) = \int_{\mathcal{X}} \frac{\partial \log p(x \mid \theta)}{\partial \theta_i} \frac{\partial \log p(x \mid \theta)}{\partial \theta_j} p(x \mid \theta) \, dx.$$

The **Jeffreys' prior** is

$$\varphi_{\Theta}(\theta) = \frac{\sqrt{\det I(\theta)}}{\int_{\Theta} \sqrt{\det I(\theta')} \, d\theta'}.$$

Example 7.6 For $p = 1$, with log-likelihood $L_\Theta := \log\{\theta^k(1-\theta)^{n-k}\}$,

$$I(\theta) = \mathbb{E}_X\left[\left(\frac{\partial L_\Theta}{\partial \theta}\right)^2\right] = \mathbb{E}_X\left[\left\{\frac{k-n\theta}{\theta(1-\theta)}\right\}^2\right] = \frac{n}{\theta(1-\theta)}.$$

Hence the Jeffreys' prior is

$$\varphi_\Theta(\theta) = \frac{C_\Theta}{\sqrt{\theta(1-\theta)}}, \quad C_\Theta = \left[\int_0^1 \frac{1}{\sqrt{\theta(1-\theta)}} d\theta\right]^{-1}.$$

■

We can confirm invariance by reparameterizing and obtaining the same form.

Example 7.7 Using $t = \sqrt{\theta}$ instead of θ, the log-likelihood is

$$L_T := \log\{t^{2k}(1-t^2)^{n-k}\}.$$

Then the Jeffreys' prior is

$$\varphi_T(t) = \frac{C_T}{\sqrt{1-t^2}}, \quad C_T = \left[\int_0^1 \frac{1}{\sqrt{1-t^2}} dt\right]^{-1} \tag{7.11}$$

(Problem 70). If $C_T = 2C_\Theta$, then

$$\varphi_\Theta(\theta)\,d\theta = \frac{C_\Theta}{\sqrt{\theta(1-\theta)}} d\theta = \frac{C_\Theta}{\sqrt{t^2(1-t^2)}} \frac{d\theta}{dt} dt = \frac{C_T}{\sqrt{1-t^2}} dt = \varphi_T(t)\,dt.$$

■

Thus, Jeffreys' prior is invariant to reparameterization and preserves noninformativeness regardless of representation; in this sense it is a "truly noninformative prior."

Let us consider a Beta prior $Be(a, b)$ with $a, b > 0$, covering not only $a = b = 1$ (uniform) but also $a = b = 0.5$ (Jeffreys' prior).

$$\begin{aligned} q(x_1, \ldots, x_n) &:= \int_0^1 \theta^k(1-\theta)^{n-k}\varphi(\theta)d\theta = \int_0^1 \frac{\theta^{k+a-1}(1-\theta)^{n-k+b-1}}{B(a,b)} d\theta \\ &= \frac{B(k+a, n-k+b)}{B(a,b)} \end{aligned} \tag{7.12}$$

Using (2.11), we obtain

$$q(x_1, \ldots, x_n) = \frac{\Gamma(k+a)\Gamma(n-k+b)\Gamma(a+b)}{\Gamma(n+a+b)\Gamma(a)\Gamma(b)}. \tag{7.13}$$

In particular, for the uniform prior ($a = b = 1$),

$$q(x_1, \ldots, x_n) = \frac{k!(n-k)!}{(n+1)!}. \tag{7.14}$$

Next, given data $x_1, \ldots, x_n$, the predictive probabilities for $x_{n+1} = 0, 1$ are

$$q(x_{n+1} \mid x_1, \ldots, x_n) := \frac{q(x_1, \ldots, x_n, x_{n+1})}{q(x_1, \ldots, x_n)} = \begin{cases} \dfrac{k+a}{n+a+b}, & x_{n+1} = 1, \\ \dfrac{n-k+b}{n+a+b}, & x_{n+1} = 0 \end{cases} \tag{7.15}$$

(Problem 72). Indeed, from (2.8) and (7.13),

$$\frac{q(x_1, \ldots, x_n, 1)}{q(x_1, \ldots, x_n)} = \frac{\Gamma(k+1+a)\Gamma(n-k+b)}{\Gamma(n+1+a+b)} / \frac{\Gamma(k+a)\Gamma(n-k+b)}{\Gamma(n+a+b)} = \frac{k+a}{n+a+b},$$
$$\frac{q(x_1, \ldots, x_n, 0)}{q(x_1, \ldots, x_n)} = \frac{\Gamma(k+a)\Gamma(n-k+1+b)}{\Gamma(n+1+a+b)} / \frac{\Gamma(k+a)\Gamma(n-k+b)}{\Gamma(n+a+b)} = \frac{n-k+b}{n+a+b}.$$

Hence the marginal likelihood can be written as a product of these conditionals; for $x_i \in \{0, 1\}$,

$$\begin{aligned} q(x_1, \ldots, x_n) &= \prod_{i=1}^{n} q(x_i \mid x_1, \ldots, x_{i-1}) \\ &= \prod_{i=1}^{n} \left\{ \left(\frac{k_{i-1}+a}{i-1+a+b} \right)^{x_i} \left(\frac{i-1-k_{i-1}+b}{i-1+a+b} \right)^{1-x_i} \right\}, \end{aligned}$$

where k_{i-1} is the frequency of 1's among $x_1, \ldots, x_{i-1}$.

Example 7.8 For $n = 3$, the values of $q(x_1, x_2, x_3)$, $x_1, x_2, x_3 \in \{0, 1\}$ are as follows: For $a = b = 0.5$,

$$q(0, 0, 0) = \frac{\frac{1}{2}}{\frac{1}{2}+\frac{1}{2}} \cdot \frac{1+\frac{1}{2}}{1+\frac{1}{2}+\frac{1}{2}} \cdot \frac{2+\frac{1}{2}}{2+\frac{1}{2}+\frac{1}{2}} = \frac{5}{16},$$

$$q(0, 0, 1) = \frac{\frac{1}{2}}{\frac{1}{2}+\frac{1}{2}} \cdot \frac{1+\frac{1}{2}}{1+\frac{1}{2}+\frac{1}{2}} \cdot \frac{\frac{1}{2}}{2+\frac{1}{2}+\frac{1}{2}} = \frac{1}{16};$$

for $a = b = 1$,

$$q(0, 0, 0) = \frac{1}{1+1} \cdot \frac{1+1}{1+1+1} \cdot \frac{2+1}{2+1+1} = \frac{1}{4},$$

$$q(0, 0, 1) = \frac{1}{1+1} \cdot \frac{1+1}{1+1+1} \cdot \frac{1}{2+1+1} = \frac{1}{12}.$$

By symmetry, the other six values follow similarly. The $a = b = 1$ case is less sensitive to the specific data values than $a = b = 0.5$. ■

If we know 1's are more frequent, we can set $a > b$. Since $a, b > 0$ parameterize the prior φ and can be interpreted as (possibly non-integer) pseudo-counts, we call them **prior frequencies**.

When $X_1, \ldots, X_n$ are i.i.d. and take α distinct values $1, \ldots, \alpha$, the results extend as follows.

Proposition 7.1 *Suppose X takes α values $1, \ldots, \alpha$ with frequencies $k_1, \ldots, k_\alpha$, and prior frequencies $a_1, \ldots, a_\alpha > 0$. Then the marginal likelihood of $X = x_1, \ldots, x_n$ is*

$$q(x_1, \ldots, x_n) = \frac{\prod_{j=1}^{\alpha} \Gamma(k_j + a_j)}{\Gamma(\sum_{j=1}^{\alpha}(k_j + a_j))} \cdot \frac{\Gamma(\sum_{j=1}^{\alpha} a_j)}{\prod_{j=1}^{\alpha} \Gamma(a_j)}. \tag{7.16}$$

Proof The marginal likelihood for $X = 1$ versus $2 \leq X \leq \alpha$ is

$$\frac{B(k_1 + a_1, \sum_{h=2}^{\alpha}(k_h + a_h))}{B(a_1, \sum_{h=2}^{\alpha} a_h)} = \frac{\Gamma(k_1 + a_1)\Gamma(\sum_{h=2}^{\alpha}(k_h + a_h))}{\Gamma(\sum_{j=1}^{\alpha}(k_j + a_j))} \cdot \frac{\Gamma(\sum_{h=1}^{\alpha} a_h)}{\Gamma(a_1)\Gamma(\sum_{h=2}^{\alpha} a_h)}.$$

Given $X \geq 2$, the marginal likelihood for $X = 2$ versus $X \geq 3$ is

$$\frac{B((k_2 + a_2), \sum_{h=3}^{\alpha}(k_h + a_h))}{B(a_2, \sum_{h=3}^{\alpha} a_h)} = \frac{\Gamma(k_2 + a_2)\Gamma(\sum_{h=3}^{\alpha}(k_h + a_h))}{\Gamma(\sum_{j=2}^{\alpha}(k_j + a_j))} \cdot \frac{\Gamma(\sum_{h=2}^{\alpha} a_h)}{\Gamma(a_2)\Gamma(\sum_{h=3}^{\alpha} a_h)}.$$

Multiplying these $\alpha - 1$ terms,

$$\prod_{j=2}^{\alpha} \frac{B((k_{j-1} + a_{j-1}), \sum_{h=j}^{\alpha}(k_h + a_h))}{B(a_{j-1}, \sum_{h=j}^{\alpha} a_h)} \tag{7.17}$$

yields (7.16) (Problem 75). ■

Using (7.17), we can implement marginal likelihood in Python. Since raw values can be tiny, we output the negative log marginal likelihood. The vector **k** holds counts `k[0]`,...,`k[m-1]`, and a holds prior frequencies.

```
from scipy.special import gammaln  # log Γ
from scipy.stats import multinomial
```

```
def Q_1(k, a):
    k = np.asarray(k, dtype=float)
    a = np.asarray(a, dtype=float)
    m = len(k)
    log_q = 0.0
    for j in range(1, m):  # R: 2:m
        num = (gammaln(k[j-1] + a[j-1]) + gammaln(np.sum(k[j:] + a[j:]))
               - gammaln(k[j-1] + a[j-1] + np.sum(k[j:] + a[j:])))
        den = (gammaln(a[j-1]) + gammaln(np.sum(a[j:]))
               - gammaln(a[j-1] + np.sum(a[j:])))
        log_q += (num - den)
    return -log_q
```

Example:

```
from scipy.stats import multivariate_normal
n = 100
m = 4
prob = np.ones(m) / m
k = multinomial.rvs(n, prob)
a = np.ones(m)
print(Q_1(k, a))
```

If X and Y take l and m values, respectively, and we wish to compute the marginal likelihood for

$$(X, Y) = (x_1, y_1), \ldots, (x_n, y_n),$$

we can treat $Z = (X, Y)$ as a variable with lm categories and proceed as above. In practice, "use `l` levels for `x` and `m` for `y`, form `idx = xi + l*yi`, and count via `np.bincount`."

Define:

```
def joint_counts_vector(x, y):
    x = np.asarray(x)
    y = np.asarray(y)
    levels_x, xi = np.unique(x, return_inverse=True)  # xi {0,...,l-1}
    levels_y, yi = np.unique(y, return_inverse=True)  # yi {0,...,m-1}
    l = len(levels_x); m = len(levels_y)
    idx = xi + l * yi                              # column-major like R's as.vector
    k_vec = np.bincount(idx, minlength=l*m)
    return k_vec, levels_x, levels_y
```

Example:

```
rng = np.random.default_rng(0)
n = 100
x = rng.binomial(1, 0.5, size=n)
y = rng.binomial(1, 0.25, size=n)
k, lx, ly = joint_counts_vector(x, y)    # length-4 vector
a = np.full(4, 0.5)                       # prior frequencies (a_j = 0.5)
print("k =", k.tolist())
print("Q.1(k, a) =", Q_1(k, a))
```

In general, closed forms for marginal likelihood are not available for arbitrary priors. Typically one uses **Markov chain Monte Carlo** (MCMC) to sample parameters from the posterior

$$p(\theta \mid x_1, \ldots, x_n) := \frac{\prod_{i=1}^n p(x_i \mid \theta)\varphi(\theta)}{\int_\Theta \prod_{i=1}^n p(x_i \mid \theta)\varphi(\theta)d\theta},$$

generate $\theta_1, \ldots, \theta_m \sim p(\cdot \mid x_1, \ldots, x_n)$, and approximate

$$\int_\Theta \left\{\prod_{i=1}^n p(x_i \mid \theta)\right\} p(\theta \mid x_1, \ldots, x_n)d\theta \approx \frac{1}{m}\sum_{j=1}^m \left\{\prod_{i=1}^n p(x_i \mid \theta_j)\right\}.$$

Analytic formulas, as used in this chapter, are generally limited to **exponential families** with **conjugate priors**.

7.3 Marginal Likelihood (Continuous)

Next, suppose $x_1, \ldots, x_n \in \mathbb{R}^p$ are i.i.d. from a multivariate normal with unknown mean $\mu \in \mathbb{R}^p$ and unknown covariance $\Sigma \in \mathbb{R}^{p\times p}$.

Let $\kappa_0, \nu_0 > p-1$, $\mu_0 \in \mathbb{R}^p$, $\Lambda_0 \in \mathbb{R}^{p\times p}$, and set the prior as

$$\mu \mid \Sigma \sim N(\mu_0, \Sigma/\kappa_0), \qquad \Sigma \sim W^{-1}(\Lambda_0, \nu_0), \tag{7.18}$$

where $W^{-1}(\Lambda_0, \nu_0)$ is the inverse-Wishart distribution with parameters (Λ_0, ν_0) as in (2.30). Set

$$\bar{x} := \frac{1}{n}\sum_{i=1}^n x_i, \qquad \kappa_n := \kappa_0 + n, \qquad \nu_n := \nu_0 + n$$

and

$$\Lambda_n := \Lambda_0 + \sum_{i=1}^n (x_i - \bar{x})(x_i - \bar{x})^T + \frac{\kappa_0 n}{\kappa_0 + n}(\bar{x} - \mu_0)(\bar{x} - \mu_0)^T.$$

Proposition 7.2 (K. Murphy [18]) *For $x_1, \ldots, x_n$ i.i.d. from $N(\mu, \Sigma)$ with unknown $\mu \in \mathbb{R}^p$ and $\Sigma \in \mathbb{R}^{p\times p}$ and prior (7.18), the marginal likelihood is*

$$\frac{1}{\pi^{np/2}} \frac{\Gamma_p(\nu_n/2)}{\Gamma_p(\nu_0/2)} \frac{\det(\Lambda_0)^{\nu_0/2}}{\det(\Lambda_n)^{\nu_n/2}} \left(\frac{\kappa_0}{\kappa_n}\right)^{p/2}.$$

Proof See the Appendix at the end of the chapter. ∎

Using an inverse-Wishart prior for Σ yields a posterior of the same family with updated parameters (conjugacy). Alternatively, one may use a Wishart prior for the precision matrix $\Lambda = \Sigma^{-1}$. Both are applicable, but priors on Σ are often more convenient.

In Python, the negative log marginal likelihood $-\log p(D)$ can be implemented as:

```
from scipy.special import multigammaln

def Q_3(x, kappa_0, nu_0, mu_0, Lambda_0):
    x = np.asarray(x, dtype=float)
    n, d = x.shape
    kappa_n = kappa_0 + n
    nu_n = nu_0 + n

    x_bar = np.mean(x, axis=0)
    Xm = x - x_bar
    S = Xm.T @ Xm

    mu_diff = (x_bar - np.asarray(mu_0, dtype=float)).reshape(d, 1)
    M = (kappa_0 * n) / (kappa_0 + n) * (mu_diff @ mu_diff.T)
    Lambda_n = np.asarray(Lambda_0, dtype=float) + S + M

    # Stable log det
    sign0, logdet0 = np.linalg.slogdet(Lambda_0)
    signn, logdetn = np.linalg.slogdet(Lambda_n)
    if sign0 <= 0 or signn <= 0:
        raise np.linalg.LinAlgError("Lambda_0 / Lambda_n must be SPD.")

    value = (n * d / 2) * np.log(np.pi) \
            - multigammaln(nu_n / 2, d) + multigammaln(nu_0 / 2, d) \
            - (nu_0 / 2) * logdet0 + (nu_n / 2) * logdetn \
            - (d / 2) * np.log(kappa_0 / kappa_n)
    return value
```

Example 7.9 We test the function Q_3 on synthetic data.

```
# Quick check
n = 100
x = multivariate_normal.rvs(mean=[0,0], cov=np.eye(2), size=n)
a = np.random.randn(2)
A = np.outer(a, a)
y = multivariate_normal.rvs(mean=[0,0], cov=A, size=n)
print(Q_3(x, 1, 2, np.array([0,0]), np.eye(2)))
print(Q_3(y, 1, 2, np.array([0,0]), np.eye(2)))
```

∎

7.4 Essence of BIC

The negative log marginal likelihood

$$F(x_1, \ldots, x_n) := -\log q(x_1, \ldots, x_n) \tag{7.19}$$

is called the *free energy*. Outside of exponential families with conjugate priors, exact values are typically unavailable and MCMC approximations are required.

We call a statistical model $p(\cdot \mid \theta), \theta \in \Theta \subseteq \mathbb{R}^d$, *regular* with respect to the true distribution $q(\cdot)$ if:

- The Fisher information matrix has rank d.
- The minimizer $\theta_* \in \Theta$ of the Kullback–Leibler divergence from q to $p(\cdot \mid \theta)$ is unique.
- θ_* is not on the boundary of Θ.

Here, the KL divergence is

$$\int_{\mathcal{X}} q(x) \log \frac{q(x)}{p(x \mid \theta)} dx,$$

nonnegative by

$$\log x \leq x - 1, \qquad x > 0, \tag{7.20}$$

and equals zero only if $p(X \mid \theta) = q(X)$ almost surely. If multiple minimizers exist, the MLE is not unique.

Under regularity, the free energy (7.19) is well approximated by BIC (7.2).

We recall **Stirling's formula** for the Gamma function:

$$\Gamma(x) \sim \sqrt{2\pi} e^{-x} x^{x-\frac{1}{2}}, \tag{7.21}$$

meaning $f(x) \sim g(x)$ if $f(x)/g(x) \to 1$ as $x \to \infty$.

Example 7.10 Using Stirling's formula in (7.16) with $a_j = 0.5$ gives

$$\Gamma(k_j + \frac{1}{2}) \sim \sqrt{2\pi} \exp\{-(k_j + \frac{1}{2})\} \left(k_j + \frac{1}{2}\right)^{k_j},$$

$$\Gamma(n + \frac{\alpha}{2}) \sim \sqrt{2\pi} \exp\{-(n + \frac{\alpha}{2})\} \left(n + \frac{\alpha}{2}\right)^{\alpha},$$

so, ignoring terms that vanish as $n \to \infty$,

$$\sum_{j=1}^{\alpha} -k_j \log \frac{k_j + 1/2}{n + \alpha/2} + \frac{\alpha - 1}{2} \log(n+1) + \frac{\alpha - 1}{2} \log(2\pi) - \log \frac{\Gamma(\alpha/2)}{\Gamma(1/2)^{\alpha}}. \tag{7.22}$$

Writing constant deviations as $O(1)$, we have

$$-\log q(x_1, \ldots, x_n) = \sum_{j=1}^{\alpha} -k_j \log \frac{k_j}{n} + \frac{\alpha - 1}{2} \log n + O(1) \tag{7.23}$$

(Problem 73). ■

Marginal likelihood, like BIC, is applicable to model selection such as linear regression. Comparing the ratio $Q(x^n, y^n, z^n)/Q(x^n, y^n)$ to $Q(x^n, z^n)/Q(x^n)$ determines whether Z should use (X, Y) or only X as regressors:

$$\frac{Q(x^n, y^n, z^n)}{Q(x^n, y^n)} > \frac{Q(x^n, z^n)}{Q(x^n)} \iff Z \text{ uses } (X, Y) \text{ as regressors.}$$

Example 7.11 For Boston's 14th variable `med` as the response, we select a subset `index` of the first 13 continuous variables (10 in total) by maximizing the difference between the (negative log) marginal likelihood including `med` and that without it (for numerical stability, we use differences of negative logs). We also compare with the unsimplified BIC value `BIC_2` for continuous data. Although the absolute values differed by up to about 10%, both selected six variables as optimal (Fig. 7.3).

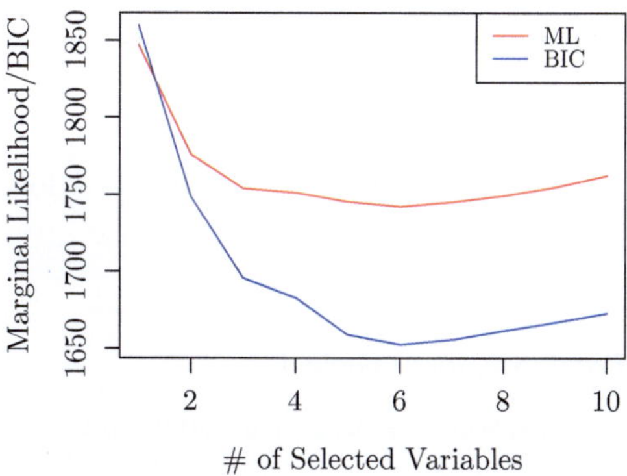

Fig. 7.3 Marginal likelihood vs. BIC on the Boston data

```
df = pd.read_csv("Boston.csv")
X = df.iloc[:, [0, 2, 4, 5, 6, 7, 9, 10, 11, 12]].to_numpy()
Y = df.iloc[:, 13].to_numpy().reshape(-1, 1)
n, p = X.shape

Q3_seq = []
for k in range(1, p + 1):
    min_score = np.inf
    for comb in itertools.combinations(range(p), k):
        X_sub = X[:, comb]
        mu_0 = np.zeros(X_sub.shape[1])
        Lambda_0 = np.eye(X_sub.shape[1])
        q3_x = Q_3(X_sub, 1, X_sub.shape[1] + 2, mu_0, Lambda_0)

        X_sub_y = np.column_stack([X_sub, Y])
        mu_0_y = np.zeros(X_sub_y.shape[1])
        Lambda_0_y = np.eye(X_sub_y.shape[1])
        q3_xy = Q_3(X_sub_y, 1, X_sub_y.shape[1] + 2, mu_0_y, Lambda_0_y)

        score = q3_xy - q3_x
        if score < min_score:
            min_score = score
    Q3_seq.append(min_score)

plt.plot(range(1, p + 1), Q3_seq, "o-", color="green")
plt.xlabel("Number of selected variables")
plt.ylabel("Q3(X,Y) - Q3(X)")
plt.title("Variable selection by marginal likelihood difference Q3(X,Y)-Q3(X)
  ")
plt.grid()
plt.show()
```

More generally, Laplace's method links the two. Let

$$l(\theta) := \frac{1}{n}\sum_{i=1}^{n} \log p(x_i \mid \theta), \qquad C := -\nabla^2 l(\hat{\theta}) = \left(-\frac{\partial^2 l}{\partial\theta_i \partial\theta_j}\bigg|_{\theta=\hat{\theta}} \right).$$

Proposition 7.3 (Laplace's Method) *The difference between* $-\log q(x_1, \ldots, x_n)$ *and*

$$\sum_{i=1}^{n} -\log p(x_i \mid \hat{\theta}) + \frac{d}{2}\log n - \frac{d}{2}\log 2\pi + \frac{1}{2}\log\det C - \log(\varphi(\hat{\theta}))$$

converges in probability to 0 *as* $n \to \infty$.

Proof See, for example, [7]. ■

Thus, when the true distribution and model are regular, BIC applies. Marginal likelihood uses a specific prior, but considering that MCMC can be computationally heavy, BIC has substantial practical advantages.

7.5 Essence of AIC

Although AIC and BIC look similar by definition, their derivations differ.

Given data $x_1, \dots, x_n$ and MLE $\hat{\theta}(x_1, \dots, x_n)$, AIC can be written as

$$AIC(x_1, \dots, x_n) := -\sum_{i=1}^{n} \log p(x_i \mid \hat{\theta}(x_1, \dots, x_n)) + d.$$

The second term d is derived so that

$$\mathbb{E}_{X_1 \cdots X_n}[AIC(X_1, \dots, X_n)] = \mathbb{E}_{X_1 \cdots X_n} \mathbb{E}_X [\log p(X \mid \hat{\theta}(X_1, \dots, X_n))],$$

where $\mathbb{E}_{X_1 \cdots X_n}$ denotes averaging over training samples used to obtain $\hat{\theta}$, and $\mathbb{E}_X$ denotes averaging over test samples.

Example 7.12 (Derivation of AIC for Linear Regression) We have observed pairs $(x_1, y_1), \dots, (x_n, y_n) \in \mathbb{R}^d \times \mathbb{R}$ to estimate β and wish to predict $(x_1, z_1), \dots, (x_n, z_n)$. For a candidate estimator γ, with $X = [x_1, \dots, x_n]^\top$, we minimize the average negative log-likelihood:

$$\begin{aligned} & -\log \left[\prod_{i=1}^{n} \frac{1}{\sqrt{2\pi\sigma^2}} \exp\left\{ -\frac{(z_i - x_i \gamma)^2}{2\sigma^2} \right\} \right] \\ = & \frac{n}{2} \log(2\pi\sigma^2) + \frac{\|z - X\beta + X\beta - X\gamma\|^2}{2\sigma^2} \\ = & \frac{n}{2} \log(2\pi\sigma^2) + \frac{\|z - X\beta\|^2}{2\sigma^2} - \frac{(z - X\beta)^\top X(\gamma - \beta)}{\sigma^2} + \frac{\|X(\gamma - \beta)\|^2}{2\sigma^2}. \end{aligned}$$

Taking expectation over $Z = z_1, \dots, z_n$, the numerator of the second term becomes $n\sigma^2$ and the third term vanishes:

$$\mathbb{E}_{Z_1 \cdots Z_n} \left[-\log \left\{ \prod_{i=1}^{n} \frac{1}{\sqrt{2\pi\sigma^2}} \exp\left(-\frac{(z_i - x_i \gamma)^2}{2\sigma^2} \right) \right\} \right] = \frac{n}{2} \log(2\pi\sigma^2 e) + \frac{\|X(\gamma - \beta)\|^2}{2\sigma^2}. \tag{7.24}$$

Restricting γ to unbiased estimators of β, for any unbiased $\tilde{\beta}$, the least squares estimator $\hat{\beta}$ satisfies

$$\mathbb{E}_{Y_1 \cdots Y_n} \left[\|X(\tilde{\beta} - \beta)\|^2 \right] \geq \mathbb{E}_{Y_1 \cdots Y_n} \left[\|X(\hat{\beta} - \beta)\|^2 \right] = d\sigma^2. \tag{7.25}$$

Substituting the RHS of (7.25) into (7.24) yields

$$\frac{n}{2}\log(2\pi\sigma^2 e) + \frac{d}{2}, \tag{7.26}$$

which AIC seeks to minimize. However, $\log\sigma^2$ is unknown; replacing with the MLE $\log\hat{\sigma}^2$ makes it smaller on average by $\frac{d+1}{n}$:

$$\mathbb{E}_{Y_1\cdots Y_n}[\log\hat{\sigma}^2] = \log\sigma^2 - \frac{d+1}{n} + O(\frac{1}{n^2}). \tag{7.27}$$

Thus,

$$\mathbb{E}_{Y_1\cdots Y_n}\Big[\frac{n}{2}\log(2\pi\hat{\sigma}^2 e) + \frac{d}{2} + \frac{d+1}{2} + O(\frac{1}{n})\Big]$$

matches (7.26) up to $O(1)$. In practice, AIC is often written as

$$AIC := n\log(\hat{\sigma}^2) + 2d.$$

For proofs of (7.25) and (7.27), see [31, 32], Chapter 5. ■

Appendix

Proof of Proposition 7.2

The marginal likelihood formula coincides with (5.29) of [18], where the derivation is omitted. We provide an elementary proof via Bayes' theorem.

Given observations $x_1, \ldots, x_n \in \mathbb{R}^p$, the likelihood $p(D \mid \mu, \Sigma)$ is

$$(2\pi)^{-\frac{np}{2}}\det(\Sigma)^{-\frac{n}{2}}\exp\left\{-\frac{1}{2}\sum_{i=1}^{n}(x_i-\mu)^\top\Sigma^{-1}(x_i-\mu)\right\} \tag{7.28}$$

and the prior for $\mu|\Sigma$ is

$$f_{\mu_0,\kappa_0}(\mu \mid \Sigma) = \left(\frac{\kappa_0}{2\pi}\right)^{p/2}\det(\Sigma)^{-1/2}\exp\left\{-\frac{1}{2}(\mu-\mu_0)^\top(\Sigma/\kappa_0)^{-1}(\mu-\mu_0)\right\},$$

with Σ having the inverse-Wishart prior (2.31) with $\nu = \nu_0$, $\Lambda = \Lambda_0$. Thus the joint prior is

$$p(\mu, \Sigma) := f_{\mu_0,\kappa_0}(\mu \mid \Sigma) f_{\nu_0,\Lambda_0}(\Sigma)$$
$$= \frac{1}{Z_0} \det(\Sigma)^{-\left(\frac{\nu_0+p}{2}+1\right)} \exp\left\{-\frac{1}{2}\mathrm{tr}\left(\Lambda_0 \Sigma^{-1}\right) - \frac{\kappa_0}{2}(\mu - \mu_0)^\top \Sigma^{-1}(\mu - \mu_0)\right\}, \tag{7.29}$$

$$Z_0 = \frac{2^{\frac{\nu_0 p}{2}} \Gamma_p\left(\frac{\nu_0}{2}\right)\left(2\pi\kappa_0^{-1}\right)^{\frac{p}{2}}}{\det(\Lambda_0)^{\frac{\nu_0}{2}}}.$$

We write this as $NIW(\kappa_0, \mu_0, \Lambda_0, \nu_0)$. The product of (7.28) and (7.29), by Bayes' theorem

$$p(\mu, \Sigma \mid D) = \frac{p(\mu, \Sigma)p(D \mid \mu, \Sigma)}{p(D)},$$

equals the product of the marginal likelihood $p(D)$ and the posterior $p(\mu, \Sigma \mid D)$. The exponent is

$$-\frac{1}{2}\mathrm{tr}(\Sigma^{-1}\Lambda_0) - \frac{1}{2}\sum_{i=1}^{n}(x_i - \mu)^T \Sigma^{-1}(x_i - \mu) - \frac{\kappa_0}{2}(\mu - \mu_0)^T \Sigma^{-1}(\mu - \mu_0)$$
$$= -\frac{1}{2}\mathrm{tr}(\Sigma^{-1}\Lambda_0) - \frac{\kappa_0 + n}{2}\mu^\top \Sigma^{-1}\mu + (\kappa_0\mu_0 + n\bar{x})^\top \Sigma^{-1}\mu - \frac{\kappa_0}{2}\mu_0^\top \Sigma^{-1}\mu_0$$
$$-\frac{1}{2}\sum_{i=1}^{n} x_i^\top \Sigma^{-1} x_i$$
$$= -\frac{1}{2}\mathrm{tr}(\Sigma^{-1}\Lambda_0) - \frac{\kappa_0 + n}{2}\left(\mu - \frac{\kappa_0\mu_0 + n\bar{x}}{\kappa_0 + n}\right)^\top \Sigma^{-1}\left(\mu - \frac{\kappa_0\mu_0 + n\bar{x}}{\kappa_0 + n}\right)$$
$$-\frac{n\kappa_0}{2(\kappa_0 + n)}(\mu_0 - \bar{x})^\top \Sigma^{-1}(\mu_0 - \bar{x}) - \frac{1}{2}\sum_{i=1}^{n}(x_i - \bar{x})^\top \Sigma^{-1}(x_i - \bar{x})$$
$$= -\frac{1}{2}\mathrm{tr}\left(\Sigma^{-1}\left\{\sum_{i=1}^{n}(x_i - \bar{x})(x_i - \bar{x})^T + \Lambda_0 + \frac{n\kappa_0}{2(\kappa_0 + n)}(\mu_0 - \bar{x})(\mu_0 - \bar{x})^\top\right\}\right)$$
$$-\frac{\kappa_0 + n}{2}\left(\mu - \frac{\kappa_0\mu_0 + n\bar{x}}{\kappa_0 + n}\right)^\top \Sigma^{-1}\left(\mu - \frac{\kappa_0\mu_0 + n\bar{x}}{\kappa_0 + n}\right)$$
$$= -\frac{1}{2}\mathrm{tr}(\Sigma^{-1}\Lambda_n) - \frac{\kappa_n}{2}(\mu - \mu_n)^\top \Sigma^{-1}(\mu - \mu_n),$$

with $\mu_n := \dfrac{\kappa_0\mu_0 + n\bar{x}}{\kappa_0 + n}$ (Problem 78); for the third equality we used $\mathrm{tr}(AB) = \mathrm{tr}(BA)$ when both are defined (Problem 41).

The non-exponential factors are

$$(2\pi)^{-\frac{np}{2}} \det(\Sigma)^{-\frac{n}{2}} \cdot \frac{1}{Z_0} \det(\Sigma)^{-\left(\frac{\nu_0+p}{2}+1\right)} = (2\pi)^{-\frac{np}{2}} \frac{Z_n}{Z_0} \cdot \frac{1}{Z_n} \det(\Sigma)^{-\left(\frac{\nu_n+p}{2}+1\right)},$$

where

$$Z_n = \frac{2^{\frac{\nu_n p}{2}} \Gamma_p\left(\frac{\nu_n}{2}\right) \left(2\pi\kappa_n^{-1}\right)^{\frac{p}{2}}}{\det(\Lambda_n)^{\frac{\nu_n}{2}}}.$$

Therefore the overall expression is

$$(2\pi)^{-\frac{np}{2}} \frac{Z_n}{Z_0} \cdot \underbrace{\frac{1}{Z_n} \det(\Sigma)^{-\left(\frac{\nu_n+p}{2}+1\right)} \exp\left\{-\frac{1}{2}\mathrm{tr}(\Sigma^{-1}\Lambda_n) - \frac{\kappa_n}{2}(\mu-\mu_n)^\top \Sigma^{-1}(\mu-\mu_n)\right\}}_{\text{pdf of } NIW(\kappa_n,\mu_n,\Lambda_n,\nu_n)}.$$

Integrating the underbraced posterior over (μ, Σ) gives 1, so the marginal likelihood is $(2\pi)^{-\frac{np}{2}} \frac{Z_n}{Z_0}$. Computing this yields

$$\frac{1}{(2\pi)^{np/2}} \frac{2^{\nu_n p/2}}{2^{\nu_0 p/2}} \frac{(2\pi/\kappa_n)^{p/2}}{(2\pi/\kappa_0)^{p/2}} \frac{\Gamma_p(\nu_n/2)}{\Gamma_p(\nu_0/2)} \frac{\det(\Lambda_0)^{\nu_0/2}}{\det(\Lambda_n)^{\nu_n/2}},$$

which proves the proposition. ■

Problems 70–78

70. Prove (7.11).
71. Show that, in Example 7.8, the marginal likelihood values for the case $a = b = 1$ agree with those computed by (7.14).
72. Write a Python program that generates 100 binary sequences of length 50 (with success probability p), and for each sequence plots $q(1 \mid x_1, \ldots, x_n)$ defined in (7.15) for $0 \le n \le 49$ on the same graph (connect points with line segments). Produce a total of four plots for the combinations $p = 0.25$, $p = 0.5$ and $a = b = 0.5$, $a = b = 1$. Describe the characteristics you observe across these plots.

73. Prove (7.22). Then, using the inequalities

$$0 \le \sum_{j=1}^{\alpha} k_j \log\left(k_j + \frac{1}{2}\right) - \sum_{j=1}^{\alpha} k_j \log k_j = \sum_{j=1}^{\alpha} k_j \log\left(1 + \frac{1}{2k_j}\right) \le \sum_{j=1}^{\alpha} k_j \cdot \frac{1}{2k_j} = \frac{\alpha}{2}$$

$$0 \le \sum_{j=1}^{\alpha} k_j \log\left(n + \frac{\alpha}{2}\right) - \sum_{j=1}^{\alpha} k_j \log n = \sum_{j=1}^{\alpha} k_j \log\left(1 + \frac{\alpha}{2n}\right) \le \sum_{j=1}^{\alpha} k_j \cdot \frac{\alpha}{2n} = \frac{\alpha}{2},$$

derive (7.23).

74. Below is a routine that finds the minimum negative log marginal likelihood. Using this, modify the program in Example 7.3 and plot a figure analogous to Fig. 7.3.

```
# Required libraries
import numpy as np
import itertools

Q_seq = []
for k in range(1, p + 1):
    Q_min = np.inf
    # Combinations of candidate columns 0..p-1 (size k)
    combos = list(itertools.combinations(range(p), k))
    m = len(combos)
    for j in range(m):
        # Prepend intercept column (index 0), then selected columns +1
        idx = combos[j]
        q = [0] + [i + 1 for i in idx]
        # Compute residual variance (not used directly but kept for
          fidelity)
        S = sigma2(X_full[:, q], y)
        Q = Q_3(X_full[:, q], y, np.zeros(k + 1), np.eye(k + 1), 1, 0.5)
        if Q < Q_min:
            Q_min = Q
    Q_seq.append(Q_min)

# Then plot Q_seq similarly to Fig. 3-10, e.g.:
# import matplotlib.pyplot as plt
# plt.plot(range(1, p+1), Q_seq, marker="o")
# plt.xlabel("Number of selected variables"); plt.ylabel("Minimum -log
  marginal likelihood")
# plt.grid(True); plt.show()
```

75. Derive (7.16) from (7.17).

76. Stirling's formula: With $C = \sqrt{2\pi}$, $Cx^{x-\frac{1}{2}}e^{-x} \le \Gamma(x) \le Cx^{x-\frac{1}{2}}e^{-x+\frac{1}{12x}}$ is known. Define the three functions below in your program and verify the inequalities:

```
import numpy as np
import matplotlib.pyplot as plt
from scipy.special import gammaln  # log Gamma
C = np.sqrt(2 * np.pi)
def f(x):  # Lower bound: log(C x^{x-1/2} e^{-x}) = log C + (x-1/2)log x
    - x
    x = np.asarray(x, dtype=float)
    return np.log(C) + (x - 0.5) * np.log(x) - x
def g(x):  # True log Gamma(x)
    x = np.asarray(x, dtype=float)
    return gammaln(x)
def h(x):  # Upper bound: log(C x^{x-1/2} e^{-x + 1/(12x)})
    x = np.asarray(x, dtype=float)
    return np.log(C) + (x - 0.5) * np.log(x) - x + 1.0 / (12.0 * x)
# Visualization for x  [3,5]
xs = np.linspace(3, 5, 400)
plt.figure(figsize=(7,4))
plt.plot(xs, g(xs) - f(xs), color="red",  lw=2, label="g(x)␣-␣f(x)")
plt.plot(xs, h(xs) - g(xs), color="blue", lw=2, ls="--", label="h(x)␣-␣g
   (x)")
plt.axhline(0, color="black", lw=1)
plt.xlim(3,5); plt.ylim(0, 0.03)
plt.xlabel("x"); plt.ylabel("log␣Gamma(x)")
plt.title("Stirling␣approximation␣vs.␣log␣Gamma(x)␣(zoomed)")
plt.legend(); plt.grid(True); plt.tight_layout(); plt.show()
```

77. Construct a routine to compute AIC. Fill in the blanks, run it, and similarly construct and run the routine for BIC.

```
import numpy as np
def AIC_value(X, y):
    """
␣␣␣␣X␣:␣(n,␣k)␣matrix␣of␣predictors
␣␣␣␣y␣:␣(n,)␣response␣vector
␣␣␣␣"""
    y = np.asarray(y).reshape(-1)
    X = np.asarray(X)
    n = y.shape[0]
    k = 0 if X.ndim == 1 and X.size == 0 else (0 if X.size == 0 else X.
       shape[1])
    if k == 0:
        y_bar = np.mean(y)
        S = np.sum((y - y_bar)**2) / n
    else:
        # Add intercept (R's lm adds it automatically; we add it
           explicitly)
        X_aug = np.column_stack([np.ones(n), X])
        beta, *_ = np.linalg.lstsq(X_aug, y, rcond=None)
        y_hat = X_aug @ beta
        S = np.sum((y_hat - y)**2) / n
    return n * np.log(S) + 2 * (k + 1)  # <-- fill-in
```

78. In the proof of Proposition 7.2, show the second equality in the derivation of the exponent using the identities below:

$$\begin{cases} a\mu^\top \Sigma^{-1}\mu + b^\top \Sigma^{-1}\mu = a\left(\mu + \frac{b}{2a}\right)^\top \Sigma^{-1}\left(\mu + \frac{b}{2a}\right) - \frac{1}{4a}b^\top \Sigma^{-1} b \\ a = -\frac{\kappa_0 + n}{2}, \quad b = n\bar{x} + \kappa_0\mu_0 \\ -\frac{1}{4a}b^\top \Sigma^{-1} b = \frac{1}{2(\kappa_0 + n)}(n\bar{x} + \kappa_0\mu_0)^\top \Sigma^{-1}(n\bar{x} + \kappa_0\mu_0) \end{cases}$$

$$\begin{cases} \frac{1}{2(\kappa_0 + n)}(n\bar{x} + \kappa_0\mu_0)^\top \Sigma^{-1}(n\bar{x} + \kappa_0\mu_0) - \frac{\kappa_0}{2}\mu_0^\top \Sigma^{-1}\mu_0 \\ = a_0\mu_0^\top \Sigma^{-1}\mu_0 + b_0^\top \Sigma^{-1}\mu_0 + c_0 \\ = a_0\left(\mu_0 + \frac{b_0}{2a_0}\right)^\top \Sigma^{-1}\left(\mu_0 + \frac{b_0}{2a_0}\right) + c_0 - \frac{1}{4a_0}b_0^\top \Sigma^{-1} b_0 \\ a_0 = -\frac{n\kappa_0}{2(\kappa_0 + n)}, \quad b_0 = \frac{n\kappa_0}{\kappa_0 + n}\bar{x}, \quad c_0 = \frac{n^2}{2(\kappa_0 + n)}\bar{x}^\top \Sigma^{-1}\bar{x} \\ c_0 - \frac{1}{4a_0}b_0^\top \Sigma^{-1} b_0 = \frac{n}{2}\bar{x}^\top \Sigma^{-1}\bar{x} \end{cases}$$

$$\frac{n}{2}\bar{x}^\top \Sigma^{-1}\bar{x} - \frac{1}{2}\sum_{i=1}^{n} x_i^\top \Sigma^{-1} x_i = -\frac{1}{2}\sum_{i=1}^{n} (x_i - \bar{x})^\top \Sigma^{-1} (x_i - \bar{x}).$$

Chapter 8
Score-Based Structure Learning

We begin by introducing structure discovery via hill climbing using the Python library `pgmpy`. With the Alarm dataset as a running example, we visualize how graphs are constructed based on scores (such as BIC).

Next, assuming the variable ordering is known, we show how to determine the parent set of each variable sequentially using information criteria (AIC and BIC). Within this framework, we explain—via the notion of boundary DAGs—why the graph constructed in this way is a Bayesian network, and we also provide a proof.

In the latter part, we introduce scores based on the marginal likelihood and discuss how they can be applied to tests of independence and conditional independence. In particular, by comparing scores using Jeffreys' prior with the BDeu score, we highlight a practical caveat: the lack of regularity in BDeu.

Finally, we present a method for learning forest structures that extends the Chow–Liu algorithm, and we show that replacing the conventional mutual information estimator with a new estimator can potentially improve the accuracy of independence testing.

8.1 Overview of Score-Based Structure Learning

From n i.i.d. observations, we study the problem of learning the structure (edge set) of a graphical model $G = (V, E)$ that represents dependencies among p random variables. Let $\mathcal{X}_j$ denote the domain of variable X_j (e.g., the reals or a finite set). The data are

$$x_i = (x_i^{(1)}, \dots, x_i^{(p)}) \in \mathcal{X}_1 \times \cdots \mathcal{X}_p, \qquad i = 1, \dots, n,$$

or equivalently the matrix $X = (x_i^{(j)}) \in (\mathcal{X}_1 \times \cdots \mathcal{X}_p)^n$.

J. Suzuki, *Graphical Models and Causal Discovery with Python*,
https://doi.org/10.1007/978-981-95-5308-2_8

Graphical models represent multiple conditional independences among variables simultaneously. One approach, therefore, is to run statistical tests and pick a structure that encodes those conditional independences accepted as true. This is called **constraint-based structure learning**. Algorithms such as the PC algorithm (Chap. 5) fall into this category.

In this chapter we consider computing a score (using information criteria or marginal likelihood) for each candidate structure and selecting the structure with the best score. Depending on whether we use AIC, BIC, or marginal likelihood, the learned structure can differ.

Whereas the PC algorithm assumed faithfulness, **score-based** structure learning does not, and as a consequence, finding a globally optimal solution can be computationally expensive. Hence, approximate solutions are used in practice.

In Python, `pgmpy` is a commonly used library for probabilistic graphical models (Bayesian and Markov networks). The Alarm dataset (discrete, $p = 37$) is a standard BN structure learning benchmark. To build intuition for this chapter, we first examine how a **hill climbing** procedure searches for an approximate solution.

Hill climbing starts from an initial BN structure and iteratively applies the single edit—edge insertion, deletion, or reversal—that yields the largest score improvement. The algorithm terminates when no further improvement is possible, returning the current structure as (locally) optimal.

This approach is relatively fast, but because each step makes a local choice, it may fail to reach the global optimum. Moreover, the result can depend on the initialization. Since we do not know the ground truth, it is common to run the algorithm multiple times with random initializations.

```
!pip install --upgrade --force-reinstall --no-deps "pgmpy==0.1.24.post0"
!pip install --upgrade "bnlearn==0.8.9" "networkx<3" pandas numpy
```

```
import numpy as np
import bnlearn as bn #
# Load the Alarm data (DataFrame)
df = bn.import_example(data='alarm')    # -> pandas.DataFrame

# Rename columns to "1","2",...,"p"
p = df.shape[1]
df.columns = [str(i) for i in range(1, p+1)]

# Structure learning (Hill Climbing + BIC)
model = bn.structure_learning.fit(df, methodtype='hc', scoretype='bic')

# Obtain (from, to) edges (strings) -> convert to integers
edge_list = list(model['model'].edges())              # e.g., [("1","2"),
   ...]
u = np.array([int(a) for (a, b) in edge_list], int)    # parent nodes
v = np.array([int(b) for (a, b) in edge_list], int)    # child nodes
edges = np.column_stack([u, v])                        # 2-column edge array

print(p)        # number of nodes
print(edges)    # each row is [parent child]
```

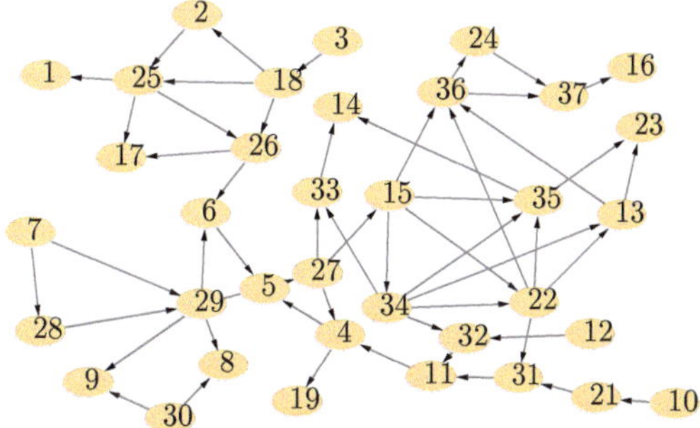

Fig. 8.1 BN learned by applying hill climbing in pgmpy to the Alarm dataset with $p = 37$. Rendered with Cytoscape

The resulting `edges` is a list of (parent, child) pairs. To visualize it, one typically uses the `networkx` library. However, with many vertices and edges the drawing can become cluttered. In Fig. 8.1, we instead render the graph with Cytoscape, a specialized tool for network visualization. For presentation or publication—especially for large networks—such dedicated tools are often preferred over `networkx`.

```
np.savetxt("alarm.csv", edges, fmt="%d", delimiter=",", header="u,v",
    comments="")
```

The file "alarm.csv" can then be imported into Cytoscape.

We construct a BN respecting a given ordering $1 < 2 < \cdots < p$: If $i < j$, then a directed edge in the DAG, when present, points from X_i to X_j. Thus, given the vertex set $V = \{1, \ldots, p\}$, for each $j = 1, \ldots, p$ we choose $\pi_j \subseteq \{1, \ldots, j-1\}$, and the directed edge set

$$E = \bigcup_{k=1}^{p} \{(j, k) \mid j \in \pi_k\}$$

is determined. We call $X_{\pi_k} = (X_j)_{j \in \pi_k}$ the **parent set** of X_k. We then select π_k so that

$$X_k \perp X_{\{1,\ldots,k-1\}-\pi_k} \mid X_{\pi_k},$$

and for any $\pi \subsetneq \pi_k$ we have

$$X_k \not\perp X_{\{1,\ldots,k-1\}-\pi} \mid X_{\pi}.$$

We refer to the resulting $G = (V, E)$, under the fixed ordering, as the **boundary DAG**. In score-based structure learning, one first fixes an ordering and then, for each vertex k, selects such a parent set π_k. Below we prove that a boundary DAG is indeed a BN.

Proposition 8.1 (Verma–Pearl [19, 37]) *Given any ordering of the variables, the corresponding boundary DAG is a Bayesian network.*

Proof See the Appendix. ■

Proposition 8.1 is one of the most important results for score-based structure learning.

In this chapter we study both cases—known and unknown variable orderings. When the ordering is known, for each $k = 2, \dots, p$ we search upstream nodes for a parent set:

$$\pi_1 = \{\}, \quad \pi_2 \subseteq \{1\}, \quad \dots, \quad \pi_k \subseteq \{1, \dots, k-1\}, \; \dots, \; \pi_p \subseteq \{1, \dots, p-1\}.$$

When the ordering is unknown, we build on the results for the known-order case.

Even with unknown orderings, we can optimize over structures using maximum posterior probability or minimum information criterion. Alternatively—at greater computational cost—we can evaluate scores across different orderings and pick the structure that optimizes the score. We discuss this latter option in Sect. 7.3. Note, however, that we cannot distinguish between Markov-equivalent structures. In practice, one may use LiNGAM (Chap. 6) to infer an ordering and then select parent sets from variables upstream.

8.2 Structure Learning via Information Criteria

In this section we assume the variable ordering is known and describe how to select parent sets using AIC and BIC [27].

When all variables are discrete, the parameters are the conditional probabilities $P(X_k = x^{(k)} \mid X_{\pi_k} = (x^{(j)})_{j \in \pi_k})$ for each configuration $s = (x^{(j)})_{j \in \pi_k}$ of the parent set X_{π_k}. If X_k takes α_k values (so the number of free parameters is $\alpha_k - 1$ due to normalization) and s ranges over $\prod_{j \in \pi_k} \alpha_j$ configurations, then the number of parameters is

$$d_k = (\alpha_k - 1) \prod_{j \in \pi_k} \alpha_j .$$

Given n observations of $(X_1, \dots, X_p)$,

$$(x_i^{(1)}, \dots, x_i^{(p)})_{i=1}^n ,$$

let n_s be the frequency of state s and $n_{h,s}$ the joint frequency of state s together with value h of X_k. The likelihood is

$$\prod_s \prod_{h=1}^{\alpha_k} \left(\frac{n_{h,s}}{n_s} \right)^{n_{h,s}} ,$$

with $\sum_s n_s = n$ and $\sum_{h=1}^{\alpha_k} n_{h,s} = n_s$. The negative log-likelihood, i.e., the **empirical entropy**, is

$$H_k = -\sum_s \sum_{h=1}^{\alpha_k} n_{h,s} \log \frac{n_{h,s}}{n_s}.$$

Hence,

$$AIC := H_k + d_k, \tag{8.1}$$

$$BIC := H_k + \frac{d_k}{2} \log n, \tag{8.2}$$

and the parent set π_k is chosen to minimize these scores. Repeat for $k = 2, 3, \ldots, p$ (with $\pi_1 = \{\}$).

Example 8.1 Let $p = 3$, $n = 10$, $\alpha_1 = 2$, $\alpha_2 = 2$, $\alpha_3 = 3$, and suppose we observe:

```
X1 = np.array([2, 2, 2, 2, 1, 2, 2, 2, 1, 2])
X2 = np.array([2, 2, 1, 1, 2, 1, 1, 2, 2, 2])
X3 = np.array([1, 1, 2, 2, 3, 1, 3, 1, 1, 1])
n = len(X1)
```

Example run (obtain contingency tables; see Problem 80):

```
tab_1,  _, _ = ftable_py(X1)
tab_2,  _, _ = ftable_py(X2)
tab_3,  _, _ = ftable_py(X3)
tab_12, _, _ = ftable_py(X1, X2)
tab_13, _, _ = ftable_py(X1, X3)
tab_23, _, _ = ftable_py(X2, X3)
tab_123,_, _ = ftable_py(X1, X2, X3)
print(tab_1, tab_2, tab_3)
print(tab_12, tab_13, tab_23)
print(tab_123)
# [[2 8]] [[4 6]] [[6 2 2]]
[[0 2]
 [4 4]] [[1 0 1]
 [5 2 1]] [[1 2 1]
 [5 0 1]]
[[0 0 0]
 [1 0 1]
 [1 2 1]
 [4 0 0]]
```

Example run (compute empirical entropies from the tables):

```
vals = [
    empirical_entropy(tab_1),
    empirical_entropy(tab_2),
    empirical_entropy(tab_3),
    empirical_entropy(tab_12),
    empirical_entropy(tab_13),
    empirical_entropy(tab_23),
    empirical_entropy(tab_123),
print([round(v, 6) for v in vals])
# -> [5.004024, 6.730117, 9.502705, 5.545177, 8.588343, 6.86225, 5.545177]
```

Here, `ftable_py` creates contingency tables, and `empirical_entropy` computes the empirical entropy from such tables:

```
import numpy as np, math
from typing import List, Tuple
```

```
def ftable_py(*cols: np.ndarray) -> Tuple[np.ndarray, List[np.ndarray], np.
  ndarray]:
    cols = [np.asarray(c) for c in cols]
    n = len(cols[0])
    assert all(len(c) == n for c in cols)
    if len(cols) == 1:
        levels = np.unique(cols[0])
        tab = np.zeros((1, len(levels)), dtype=int)
        for v in cols[0]:
            j = np.where(levels == v)[0][0]
            tab[0, j] += 1
        return tab, [levels], levels

    y = cols[-1]
    X_list = cols[:-1]
    X_levels = [np.unique(x) for x in X_list]
    y_levels = np.unique(y)

    m = np.prod([len(L) for L in X_levels])
    r = len(y_levels)
    tab = np.zeros((m, r), dtype=int)

    # Lexicographic index (strides) for Cartesian product
    strides = []
    s = 1
    for L in reversed(X_levels):
        strides.append(s)
        s *= len(L)
    strides = list(reversed(strides))

    for i in range(n):
        idxs = [np.where(X_levels[j] == X_list[j][i])[0][0] for j in range(
          len(X_list))]
        row = sum(idxs[j] * strides[j] for j in range(len(X_list)))
        col = np.where(y_levels == y[i])[0][0]
        tab[row, col] += 1
    return tab, X_levels, y_levels
```

```python
def empirical_entropy(tab: np.ndarray) -> float:
    h = 0.0
    m, r = tab.shape
    for i in range(m):
        ss = tab[i, :].sum()
        if ss > 0:
            for j in range(r):
                if tab[i, j] > 0:
                    h -= tab[i, j] * math.log(tab[i, j] / ss)
    return h
```

Since $0.5 \log n = 1.15$, for π_2 and π_3 we obtain:

π_2	H_2	d_2	AIC	BIC
{}	6.73	1	7.73	7.88
{1}	5.55	2	7.54	7.84

π_3	H_3	d_3	AIC	BIC
{}	9.50	2	11.50	11.80
{1}	8.59	4	12.59	13.19
{2}	6.86	4	10.86	11.46
{1, 2}	5.55	8	13.55	14.75

Thus both AIC and BIC select $\pi_2 = \{1\}$ and $\pi_3 = \{2\}$, from which we obtain the BN structure (Fig. 8.2). ■

Functions to compute AIC and BIC are as follows:

```python
def IC_discrete(X: np.ndarray, y: np.ndarray, proc: str = "AIC") -> float:
    if X.ndim == 1:
        tab, _, _ = ftable_py(X, y)  # one predictor: rows = its states
    else:
        # two or more predictors: rows = Cartesian product of their states;
          columns = states of y
        tab, _, _ = ftable_py(*[X[:, j] for j in range(X.shape[1])], y)
    n = len(y)
    m, r = tab.shape
    d = m * (r - 1)  # number of parameters (d_k in the text)
    H = empirical_entropy(tab)
    if proc.upper() == "AIC":
        return H + d
    else:
        return H + d * math.log(n) / 2.0
```

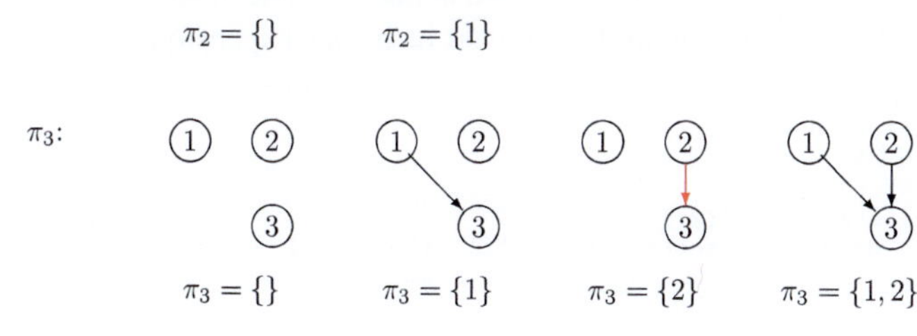

Fig. 8.2 From $\pi_1 = \{\}$, $\pi_2 = \{1\}$, and $\pi_3 = \{2\}$, the final BN $1 \to 2 \to 3$ is obtained

The continuous case is analogous. Using the functions AIC_3 or BIC_3 from Chap. 7, we can compute the parent set π_k for each $k = 2, \ldots, p$. Below is a function that enumerates all subsets of predictors and returns the set with the smallest information criterion (IC):

```
from typing import Dict, Optional, List

def IC_min(X: np.ndarray, y: np.ndarray, proc: str = "AIC") -> Dict[str,
    object]:
    y = np.asarray(y)
    n = len(y)
    if X.ndim == 1:
        X = X.reshape(-1, 1)
    p = X.shape[1]
    best_val = n * math.log(np.sum((y - y.mean()) ** 2) / n)  # constant
        model
    best_set: Optional[List[int]] = None
    if p == 0:
        return {"value": best_val, "set": best_set}

    L = 2 ** p
    for i in range(2, L + 1):          # skip the empty set (i=1)
        S1 = decimal_to_set(i - 1)     # 1-based
        cols0 = [s - 1 for s in S1]    # convert to 0-based
        X_sub = X[:, cols0] if cols0 else np.zeros((n, 0))
        val = AIC_3(X_sub, y) if proc.upper() == "AIC" else BIC_3(X_sub, y)
        if val < best_val:
            best_val, best_set = val, S1
    return {"value": best_val, "set": best_set}
```

Here, AIC_3 and BIC_3 are from Chap. 7.

```
import pandas as pd
df = pd.read_csv("Boston.csv")
# Predictors: 0,2,4,5,6,7,9,10,11,12 / Response: 13 (medv)
X = df.iloc[:, [0, 2, 4, 5, 6, 7, 9, 10, 11, 12]].to_numpy(float)
y = df.iloc[:, 13].to_numpy(float)
IC_min(X, y, proc="AIC"), IC_min(X, y, proc="BIC")
```

Example output:

```
({'value': np.float64(1655.20347628248), 'set': [1, 3, 4, 6, 8, 9, 10]},
 {'value': np.float64(1679.9220366459365), 'set': [4, 6, 8, 9, 10]})
```

The function decimal_to_set converts a nonnegative decimal integer to the subset of $\{1, \ldots, L\}$ corresponding to the 1-bits of its binary expansion:

```
def decimal_to_set(x: int) -> List[int]:
    if x == 0:
        return []
    hi = int(math.log2(x))
    return decimal_to_set(x - (1 << hi)) + [hi + 1]
```

By determining the parent sets in order, we obtain a BN:

```
p = X.shape[1]
parent = [[] for _ in range(p)]   # parent[0] is empty
for j in range(1, p):
    res = IC_min(X[:, :j], X[:, j], proc="BIC")
    # Keep 1-based indices (e.g., [1,3,4])
    parent[j] = [] if res["set"] is None else list(res["set"])
parent
```

Mathematically, the BN has now been determined. To visualize it we can use `networkx`.

```
# Parent sets -> draw directed graph
import numpy as np
import networkx as nx
import matplotlib.pyplot as plt

p = 10  # number of nodes

# Build edges: parent j  -> child i (both remain 1-based)
edges = []
for i in range(1, p + 1):           # child i = 1..p
    for j in parent[i - 1]:          # parent set (1-based)
        edges.append((j, i))         # j -> i

# Create graph
G = nx.DiGraph()
G.add_nodes_from(range(1, p + 1))
G.add_edges_from(edges)

# Optional labels: single-character labels for nodes
labels_text = ["C", "I", "N", "R", "A", "R", "T", "P", "B", "L"]  # length p
node_labels = {i + 1: labels_text[i] for i in range(p)}

# Layout: circular (akin to R's layout_with_fr alternative)
pos = nx.circular_layout(G)

# Draw
plt.figure(figsize=(6, 5))
nx.draw_networkx(
    G, pos=pos, labels=node_labels,
    with_labels=True,
    node_size=900,
    font_size=10,
    node_color="yellow",
    font_color="black",
    arrowsize=12,
    edgecolors="black",
)
plt.axis("off")
plt.title("BN structure from parent sets (Python / networkx)")
plt.show()
```

Figure 8.3 shows the resulting graphs. The BN obtained via AIC has more edges than the one via BIC.

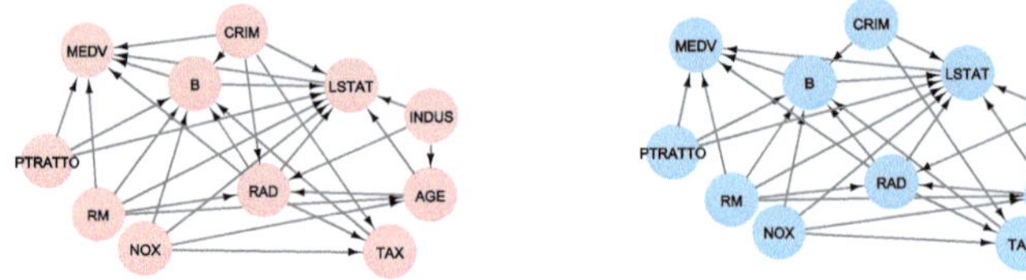

Fig. 8.3 BNs learned from the Boston dataset using AIC (left) and BIC (right)

8.3 Structure Learning via Marginal Likelihood

In this section we examine the problem of finding a BN structure that maximizes the marginal likelihood.

We begin with the case of $p = 2$ random variables. There are two possibilities for an undirected graph (independent or not) and three possibilities for a DAG (Fig. 3.1). Suppose we observe samples of random variables X, Y: $x^n := (x_1, \ldots, x_n)$ and $y^n := (y_1, \ldots, y_n)$.

For the moment, assume the variables are discrete. First, let the joint distribution of $(X, Y) = (x_i, y_i)$ for $i = 1, \ldots, n$ be written using an unknown parameter θ_{XY} as $P(X = x_i, Y = y_i \mid \theta)$. Assume also that the marginals can be written as $P(X = x_i \mid \theta)$ and $P(Y = y_i \mid \theta)$. We wish to determine whether $P(X, Y) = P(X)P(Y)$ holds (i.e., whether X and Y are independent). However, with only this information it is difficult to reach a conclusion. We therefore assume it is known that the prior probability of independence is $0 < p < 1$ and that the priors for θ_X, θ_Y, and θ_{XY} are π_X, π_Y, and π_{XY}, respectively. Suppose that, by some means, we can compute

$$Q_X(x^n) := \int_{\Theta_X} \prod_{i=1}^{n} P(X = x_i \mid \theta)\, \pi_X(\theta)\, d\theta,$$

$$Q_Y(y^n) := \int_{\Theta_Y} \prod_{i=1}^{n} P(Y = y_i \mid \theta)\, \pi_Y(\theta)\, d\theta,$$

$$Q_{XY}(x^n, y^n) := \int_{\Theta_X \times \Theta_Y} \prod_{i=1}^{n} P(X = x_i, Y = y_i \mid \theta)\, \pi_{XY}(\theta)\, d\theta,$$

where Θ_X and Θ_Y denote the domains of θ_X and θ_Y.

Then, if

$$p\, Q_X(x^n) Q_Y(y^n) \geq (1 - p)\, Q_{XY}(x^n, y^n), \tag{8.3}$$

we conclude that X and Y are independent; otherwise, we conclude they are not independent. In this case we are making the decision that maximizes the posterior under the assumed priors. The conclusion depends on the choice of the priors p, π_X,

π_Y, and π_{XY}. However, it is known that as the sample size n grows, the influence of such priors becomes negligible.

Example 8.2 We construct a procedure that, given binary sequences of length n for X and Y, decides whether they are independent or not. The function `Q_1` is the same as the one created in Chap. 7.

```
n = 100
# ===== Independent case =====
x = rng.binomial(1, 1/4, size=n)
y = rng.binomial(1, 1/2, size=n)
# z encodes (x, y): (0,0)->0, (1,0)->1, (0,1)->2, (1,1)->3
z = x + 2 * y
k_x = np.bincount(x, minlength=2)
k_y = np.bincount(y, minlength=2)
k_z = np.bincount(z, minlength=4)
result_indep = Q_1(k_x, np.full(2, 0.5)) + Q_1(k_y, np.full(2, 0.5)) \
               - Q_1(k_z, np.full(4, 0.5))
print("indep case:", result_indep)

# ===== Dependent case =====
x = rng.binomial(1, 1/2, size=n)
y = (rng.binomial(1, 1/3, size=n) + x) % 2
z = x + 2 * y
k_x = np.bincount(x, minlength=2)
k_y = np.bincount(y, minlength=2)
k_z = np.bincount(z, minlength=4)
result_dep = Q_1(k_x, np.full(2, 0.5)) + Q_1(k_y, np.full(2, 0.5)) \
             - Q_1(k_z, np.full(4, 0.5))
print("dep   case:", result_dep)
```

Since `Q_1` is the negative log marginal likelihood,

```
Q_1(k.x, rep(1/2, 2)) +
Q_1(k.y, rep(1/2, 2)) -
Q_1(k.z, rep(1/2, 4))
```

being negative is taken to indicate independence. The decision results over 500 samples for each dataset are shown in Fig. 8.4a,b. In both cases, samples plotted above the red line are judged independent. In (8.3) we set $p = 0.5$ and the prior frequency to 0.5 (Jeffreys' prior). ■

In general, following this procedure yields a correct test of whether X and Y are independent. The property that the probability of a correct decision tends to 1 as the sample size n increases is called **consistency**. We provide a proof for the multinomial case.

Proposition 8.2 *Let X and Y be random variables taking values in $\{1, \dots, \alpha\}$ and $\{1, \dots, \beta\}$, respectively (multinomial). Let x^n and y^n be sequences of length n consisting of independent realizations of X and Y, respectively. Then, with appropriate parameter choices, the rule that declares independence when*

$$-\log Q_X(x^n) - \log Q_Y(y^n) + \log Q_{XY}(x^n, y^n) \le 0,$$

and dependence otherwise, is consistent as n grows: The probability of a correct decision tends to 1.

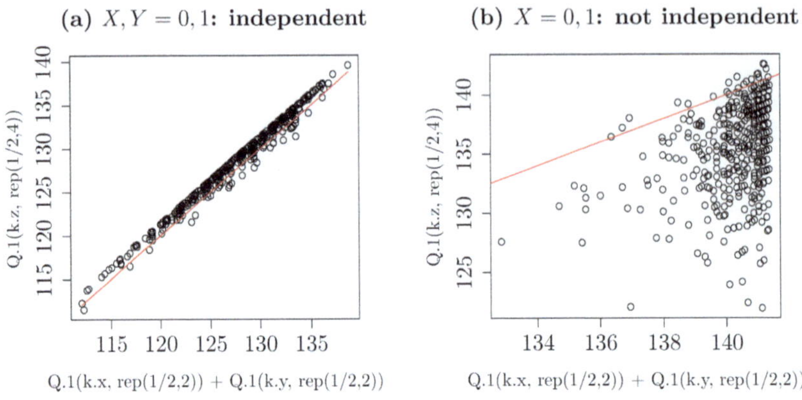

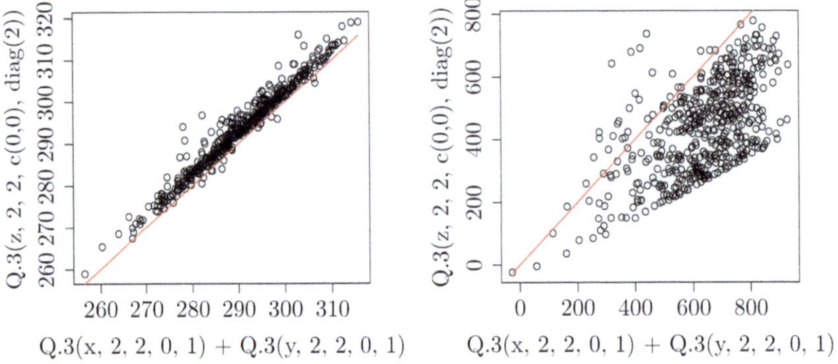

Fig. 8.4 Independence testing based on (8.3) (Example 8.2). Roughly, the independent data in (**a**) and (**c**) lie above the red line, while the dependent data in (**b**) and (**d**) lie below it

Proof First, assume $k = (k_1, \ldots, k_\alpha)$ follows a multinomial with probabilities $p = (p_1, \ldots, p_\alpha)$. Define

$$h(k, p) := \sum_{j=1}^{\alpha} -k_j \log p_j, \qquad \hat{p} = \left(\frac{k_1}{n}, \ldots, \frac{k_\alpha}{n}\right),$$

and show that

$$D(k, p) := h(k, \hat{p}) - h(k, p) = O_P(1), \tag{8.4}$$

where $O_P(1)$ denotes a term that converges in distribution as $n \to \infty$. ∎

By the mean value form of Taylor's theorem,

$$\log(1 + z) = z - \frac{z^2}{2(1 + \xi)^2}, \qquad \xi \text{ lies between } 0 \text{ and } z.$$

Applying this with $\hat{p}_j = k_j/n$, $\delta_j = \hat{p}_j - p_j$, and $z = \delta_j/p_j$ gives

$$\begin{aligned}
\mathrm{D}(k, p) &:= -n\sum_{j=1}^{\alpha} \hat{p}_j \log\frac{\hat{p}_j}{p_j} = -n\sum_{j=1}^{\alpha}(p_j+\delta_j)\left(\frac{\delta_j}{p_j} - \frac{1}{2}\frac{\delta_j^2}{p_j^2}\frac{1}{(1+\xi_j)^2}\right) \\
&= -n\sum_{j=1}^{\alpha}\left(\frac{\delta_j^2}{p_j} + \frac{\delta_j^2}{p_j}\cdot\frac{\delta_j}{p_j}\cdot\left(-\frac{1}{(1+\xi_j)^2}\right)\right) \quad (\sum_j \delta_j = 0) \\
&= -\frac{n}{2}\sum_{j=1}^{\alpha}\frac{\delta_j^2}{p_j} + R_n,
\end{aligned}$$

and by the central limit theorem we have $\delta_j = O_P(n^{-1/2})$, and with fixed α, $\sum_{j=1}^{\alpha}|\delta_j|^3 = O_P(n^{-3/2})$, hence

$$R_n := n\sum_{j=1}^{\alpha}\hat{p}_j\frac{1}{(1+\xi_j)^2}\left(\frac{\delta_j}{p_j}\right)^2\frac{\delta_j}{p_j} = O_P\left(n\sum_{j=1}^{\alpha}\frac{|\delta_j|^3}{p_j^2}\right) = O_P(n^{-1/2}) = o_P(1).$$

On the other hand,

$$n\sum_{j=1}^{\alpha}\frac{\delta_j^2}{p_j} = \sum_{j=1}^{\alpha}\frac{(k_j - np_j)^2}{np_j} \xrightarrow{d} \chi^2_{\alpha-1}$$

(Pearson's χ^2 statistic). Therefore $\mathrm{D}(k, p)$ converges in law to $-\frac{1}{2}\chi^2_{\alpha-1}$, and (8.4) follows.

Hence, from (7.23) and (8.4),

$$\begin{aligned}
-\log Q_X(x^n) &= h(k_X, \hat{p}_X) + \frac{\alpha-1}{2}\log n + O(1) \\
&= h(k_X, p_X) + \frac{\alpha-1}{2}\log n + O_P(1), \\
-\log Q_Y(y^n) &= h(k_Y, \hat{p}_Y) + \frac{\beta-1}{2}\log n + O(1) \\
&= h(k_Y, p_Y) + \frac{\beta-1}{2}\log n + O_P(1), \\
-\log Q_{XY}(x^n, y^n) &= h(k_{XY}, \hat{p}_{XY}) + \frac{\alpha\beta-1}{2}\log n + O(1) \\
&= h(k_{XY}, p_{XY}) + \frac{\alpha\beta-1}{2}\log n + O_P(1),
\end{aligned}$$

where k_X, k_Y, and k_{XY} are the observed counts for X, Y, and (X, Y) and p_X, p_Y, and p_{XY} the corresponding probabilities. If X and Y are independent, then

$$K_n := h(k_X, p_X) + h(k_Y, p_Y) - h(k_{XY}, p_{XY}) = 0$$

(Problem 84), and moreover

$$\log Q_{XY}(x^n, y^n) - \log Q_X(x^n) - \log Q_Y(y^n) = -\frac{(\alpha-1)(\beta-1)}{2}\log n + O_P(1) \tag{8.5}$$

also holds (Problem 84). Therefore, as $n \to \infty$,

$$-\log Q_X(x^n) - \log Q_Y(y^n) < -\log Q_{XY}(x^n, y^n).$$

If X and Y are not independent, then as $n \to \infty$, K_n/n converges in probability to the mutual information $I(X, Y)$, which is positive. Hence K_n grows faster than $\log n$, and thus

$$-\log Q_X(x^n) - \log Q_Y(y^n) > -\log Q_{XY}(x^n, y^n).$$

■

Proposition 8.2 guarantees that

$$J_n := \frac{1}{n}\log\frac{Q_{XY}(x^n, y^n)}{Q_X(x^n)Q_Y(y^n)} \tag{8.6}$$

is a sound measure of independence in the sense that nonpositivity indicates independence and positivity indicates dependence.

Next, we consider independence testing and its consistency for continuous variables.

Example 8.3 For a bivariate normal (X, Y), decide whether X and Y are independent from n paired observations. The function Q_3 is the same as in Chap. 7.

```
import numpy as np
rng = np.random.default_rng(0)

# ===== Dependent case =====
n = 100
a = rng.normal(size=2) * 10
A = np.outer(a, a)                    # 2x2 covariance (PSD)
A = A + 1e-8 * np.eye(2)              # small jitter for stability
z = rng.multivariate_normal([0, 0], A, size=n)
x = z[:, 0]
y = z[:, 1]
val_x  = Q_3(x,  2, 2, np.array([0.0]),          np.array([[1.0]]))
val_y  = Q_3(y,  2, 2, np.array([0.0]),          np.array([[1.0]]))
val_xy = Q_3(z,  2, 2, np.array([0.0, 0.0]),     np.eye(2))
stat = val_x + val_y - val_xy
```

```
print("stat␣(corr␣case):", stat)

# ===== Independent case (identity covariance) =====
z = rng.multivariate_normal([0, 0], np.eye(2), size=n)
x = z[:, 0]
y = z[:, 1]
val_x  = Q_3(x,  2, 2, np.array([0.0]),       np.array([[1.0]]))
val_y  = Q_3(y,  2, 2, np.array([0.0]),       np.array([[1.0]]))
val_xy = Q_3(z,  2, 2, np.array([0.0, 0.0]),  np.eye(2))
stat = val_x + val_y - val_xy
print("stat␣(indep␣case):", stat)
# negative => independent, positive => dependent
```

Since Q_3 is the negative log marginal likelihood,

```
Q_3(x, 2, 2, c(0), matrix(1, 1, 1)) +
Q_3(y, 2, 2, c(0), matrix(1, 1, 1)) -
Q_3(z, 2, 2, c(0, 0), diag(2))
```

being negative is taken to indicate independence. Results over 500 samples are shown in Fig. 8.4c,d. Again, samples plotted above the red line are judged independent. In (8.3) we set $p = 0.5$. ■

We now note the following proposition.

Proposition 8.3 *Let x^n and y^n be sequences of length n generated independently from identical distributions for X and Y, each following a normal distribution, with a prior given by (7.18). Then*

$$J_n = -\frac{1}{2}\log(1-\hat{\rho}_n^2) - \frac{1}{2n}\log n + O_P(1/n), \tag{8.7}$$

where

$$\hat{\rho}_n := \frac{\sum_{i=1}^n (x_i - \bar{x})(y_i - \bar{y})}{\sqrt{\sum_{i=1}^n (x_i - \bar{x})^2 \sum_{j=1}^n (y_j - \bar{y})^2}}$$

is the sample correlation coefficient.

Proof See the Appendix at the end of the chapter. ■

Proposition 8.4 *Under the same assumptions as Proposition 8.3, the rule based on (8.7)—nonpositive indicates independence and positive indicates dependence—is consistent as n grows.*

Proof If X and Y are independent, then $\rho = 0$, and for $X \sim N(\mu_X, \sigma_X^2)$ and $Y \sim N(\mu_Y, \sigma_Y^2)$,

$$\hat{\rho}_n - \frac{1}{n}\sum_{i=1}^n \frac{x_i - \mu_X}{\sigma_X}\frac{y_i - \mu_Y}{\sigma_Y} \to 0 \quad \text{in probability.}$$

By the central limit theorem,

$$\sqrt{n} \cdot \frac{1}{n} \sum_{i=1}^{n} \frac{x_i - \mu_X}{\sigma_X} \frac{y_i - \mu_Y}{\sigma_Y} \sim N(0, 1),$$

so $n\hat{\rho}_n^2 \sim \chi_1^2$ as $n \to \infty$. Since in general $-\frac{1}{2}z \le -\frac{1}{2}\log(1-z) \le \frac{z}{2(1-z)}$,

$$-\frac{1}{2}\hat{\rho}_n^2 \le -\frac{1}{2}\log(1-\hat{\rho}_n^2) \le \frac{\hat{\rho}_n^2}{2(1-\hat{\rho}_n^2)},$$

we have $-\frac{1}{2}\log(1-\hat{\rho}_n^2) = O_P(1/n)$. Hence in (8.7) the term $(1/2)\log n$ dominates, and the expression diverges to $-\infty$. ■

If X and Y are not independent and are normally distributed, then $\rho \neq 0$, and $\hat{\rho}_n^2$ converges in probability to $\rho^2 > 0$. Thus, in (8.7) the term $-\frac{n}{2}\log(1-\hat{\rho}_n^2)$ dominates, and the expression diverges to $+\infty$.

Although we omit the proof, Propositions 8.2 and 8.4 also extend to *conditional* independence. Namely,

$$Q_{XYZ}(x^n, y^n, z^n) + Q_Z(z^n) - Q_{XZ}(x^n, z^n) - Q_{YZ}(y^n, z^n) < 0 \iff X \perp Y \mid Z.$$

Treating Z as a constant recovers Propositions 8.2 and 8.4.

From the above, marginal likelihood can distinguish among the 11 BN structures in Fig. 3.6. For example, given n samples from X, Y, Z, x^n, y^n, and z^n, if

$$Q_{XYZ}(x^n, y^n, z^n) < \frac{Q_{XZ}(x^n, z^n) Q_{YZ}(y^n, z^n)}{Q_Z(z^n)},$$

then we may judge $X \perp Y \mid Z$, narrowing it to (a),(b),(c), or (g). Furthermore, if

$$Q_{XYZ}(x^n, y^n, z^n) < Q_{XZ}(x^n, z^n)\, Q_Y(y^n),$$

then the candidate is either (b) or (g), and so on. Including the five other cases yields Table 8.1 (Problem 85). Considering all orderings amounts to comparing all 11 structures.

Table 8.1 Structures to be compared (eight for each) when the order of X, Y, Z is fixed

Direction	(a)	(b)	(c)	(d)	(e)	(f)	(g)	(h)	(i)	(j)	(k)
$X \to Y \to Z$	✓	✓	✓	✓	✓	✓				✓	✓
$X \to Z \to Y$	✓	✓	✓	✓	✓		✓		✓		✓
$Y \to X \to Z$	✓	✓	✓	✓	✓	✓				✓	✓
$Y \to Z \to X$	✓	✓	✓	✓		✓	✓	✓			✓
$Z \to X \to Y$	✓	✓	✓	✓	✓		✓		✓		✓
$Z \to Y \to X$	✓	✓	✓	✓		✓	✓	✓			✓

That is,

$$Q_X(x^n)Q_Y(y^n)Q_Z(z^n),\ Q_X(x^n)Q_Y(y^n,z^n),\ Q_Y(y^n)Q_{ZX}(z^n,x^n),$$
$$Q_Z(z^n)Q_{XY}(x^n,y^n)$$
$$\frac{Q_{ZX}(z^n,x^n)Q_{XY}(x^n,y^n)}{Q_X(x^n)},\ \frac{Q_{XY}(x^n,y^n)Q_{YZ}(y^n,z^n)}{Q_Y(y^n)},\ \frac{Q_{ZX}(z^n,x^n)Q_{XY}(x^n,y^n)}{Q_Z(z^n)}$$
$$\frac{Q_Y(y^n)Q_Z(z^n)Q_{XYZ}(x^n,y^n,z^n)}{Q_{YZ}(y^n,z^n)},\ \frac{Q_Z(z^n)Q_X(x^n)Q_{XYZ}(x^n,y^n,z^n)}{Q_{ZX}(z^n,x^n)},$$
$$\frac{Q_X(x^n)Q_Y(y^n)Q_{XYZ}(x^n,y^n,z^n)}{Q_{XY}(x^n,y^n)},\ Q_{XYZ}(x^n,y^n,z^n) \tag{8.8}$$

and, multiplying each by a prior over structures $\pi_1, \ldots, \pi_{11} \geq 0$ with $\sum_{i=1}^{11} \pi_i = 1$, selecting the maximum yields the structure with maximum posterior probability. An analogous procedure applies to the MNs in Fig. 3.7 (Problem 86).

When using information criteria, we first fix an ordering such as $X \to Y \to Z$ and then select parent sets accordingly. Under such a fixed ordering, the arrow directions between X and Z in Fig. 3.6c and between X and Y in (f) may be reversed; nevertheless they remain Markov equivalent and receive the same score. From Table 8.1, (g), (h), and (i) are excluded (Fig. 8.5). However, if we consider all six possible orderings, we obtain a structure Markov equivalent to 1 of the 11, each expressible as a product of conditionals in the order $X \to Y \to Z$. For example, (f) and (j) can be written as

$$Q_X(x^n)\cdot\frac{Q_{XY}(x^n,y^n)}{Q_X(x^n)}\cdot\frac{Q_{YZ}(y^n,z^n)}{Q_Y(y^n)},\qquad Q_X(x^n)\cdot Q_Y(y^n)\cdot\frac{Q_{XYZ}(x^n,y^n,z^n)}{Q_{XY}(x^n,y^n)}.$$

Fig. 8.5 Eight BNs assuming the order $X \to Y \to Z$. Compared with Fig. 3.6, the directions of the arrows in **(c)** and **(f)** are reversed, but they are Markov equivalent. There are no corresponding cases for **(g)**, **(h)**, or **(i)**

8.4 When No Ordering Is Assumed

When no variable ordering is assumed, the following procedure yields the parent sets that maximize the marginal likelihood. Suppose we have n samples for each variable $X_1, \ldots, X_p$.

1. For each pair (X, S) with $X \notin S \subseteq \{X_1, \ldots, X_p\}$, compute $R(X \mid S) := \max_{U \subseteq S} Q(X \mid U)$. That is, fix X and start from $R(X \mid \{\}) = Q(X)$; as you expand S, use the recursion

$$R(X \mid S) = \max\left\{\max_{Y \in S} R\big(X \mid S \backslash \{Y\}\big),\ Q(X \mid S)\right\}$$

   to obtain $R(X \mid \cdot)$ (Fig. 8.6a).
2. Find an index ordering of $X_1, \ldots, X_p$ that maximizes the product $\prod_{i=1}^{p} R(X_i \mid \{X_1, \ldots, X_{i-1}\})$. Starting from $T(\{\}) = 1$ and expanding S, use

$$T(S) = \max_{Y \in S} T\big(S \backslash \{Y\}\big) \cdot R\big(Y \mid S \backslash \{Y\}\big).$$

   For the optimal ordering,

$$T(\{X_1, \ldots, X_p\}) = T(\{X_1, \ldots, X_{p-1}\}) \cdot R\big(X_p \mid \{X_1, \ldots, X_{p-1}\}\big) = \cdots$$
$$= \prod_{i=1}^{p} R(X_i \mid \{X_1, \ldots, X_{i-1}\})$$

   holds (Fig. 8.6b).

Finally, for each i, the set U that attains $R(X_i \mid \{X_1, \ldots, X_{i-1}\}) = Q(X_i \mid U)$ is the parent set of X_i [23, 24].

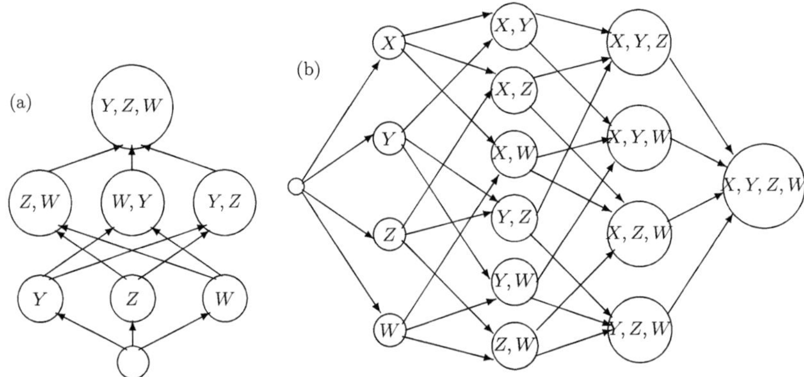

Fig. 8.6 (**a**) Procedure for computing $R(\cdot \mid \cdot)$ and (**b**) procedure for computing $T(X, Y, Z, W)$

Example 8.4 For variables X, Y, Z, W, first compute $R(X \mid \cdot)$ (Fig. 8.6a). Let $R(X) = Q(X)$. Then

$$\begin{cases} R(X \mid Y) &= \max\{R(X), Q(X \mid Y)\} = \max\{Q(X), Q(X \mid Y)\} \\ R(X \mid Z) &= \max\{R(X), Q(X \mid Z)\} = \max\{Q(X), Q(X \mid Z)\} \\ R(X \mid W) &= \max\{R(X), Q(X \mid W)\} = \max\{Q(X), Q(X \mid W)\} \end{cases}$$

$$\begin{cases} R(X \mid Z, W) &= \max\{R(X \mid Z), R(X \mid W), Q(X \mid Z, W)\} \\ R(X \mid W, Y) &= \max\{R(X \mid W), R(X \mid Y), Q(X \mid W, Y)\} \\ R(X \mid Y, Z) &= \max\{R(X \mid Y), R(X \mid Z), Q(X \mid Y, Z)\} \end{cases}$$

$$R(X \mid Y, Z, W) = \max\{R(X \mid Z, W), R(X \mid W, Y), R(X \mid Y, Z), Q(X \mid Y, Z, W)\}.$$

In the same way, compute $R(Y \mid \cdot)$, $R(Z \mid \cdot)$, and $R(W \mid \cdot)$. Next,

$$T(X) = R(X), \quad T(Y) = R(Y), \quad T(Z) = R(Z), \quad T(W) = R(W)$$

$$\begin{cases} T(X, Y) &= \max\{T(X)R(Y \mid X),\ T(Y)R(X \mid Y)\} \\ &\vdots \\ T(Z, W) &= \max\{T(Z)R(W \mid Z),\ T(W)R(Z \mid W)\} \end{cases}$$

$$\begin{cases} T(X, Y, Z) &= \max\{T(Y, Z)R(X \mid Y, Z),\ T(Z, X)R(Y \mid Z, X),\ T(X, Y)R(Z \mid X, Y)\} \\ &\vdots \\ T(Y, Z, W) &= \max\{T(Y, Z)R(W \mid Y, Z),\ T(Z, W)R(Y \mid Z, W),\ T(W, Y)R(Z \mid W, Y)\} \end{cases}$$

$$\begin{aligned} T(X, Y, Z, W) = \max \Big\{ & T(Y, Z, W)R(X \mid Y, Z, W),\ T(Z, W, X)R(Y \mid Z, W, X), \\ & T(W, X, Y)R(Z \mid W, X, Y),\ T(X, Y, Z)R(W \mid X, Y, Z) \Big\} \end{aligned}$$

(Fig. 8.6b). Since $T(X, Y, Z, W)$ factors as a product of $R(\cdot \mid \cdot)$ terms, the parent set of each variable is determined. In practice, store the maximizing parent sets while computing $T(\cdot)$, and backtrack after the overall computation finishes.

Even if you do not maximize marginal likelihood, you can replace it with minimizing AIC or BIC and run the same procedure. See Problem 87 for a concrete program.

However, even without assuming an ordering, structures that are Markov equivalent receive the same score. When edge directions must be determined, a recommended approach is to first infer an ordering via LiNGAM (Chap. 6) and then select parent sets by score among upstream variables.

8.5 BDeu

In general, the marginal likelihood is given by (7.16). In this section, let X and Y take values in $\{1, \ldots, \alpha\}$ and $\{1, \ldots, \beta\}$, respectively. Let $(x_1, y_1), \ldots, (x_n, y_n)$ be the samples, and let $n_{i,j}, n_{i,\cdot}, n_{\cdot,j}$ be the counts of $(X, Y) = (i, j)$, $X = i$, $Y = j$. For some $\delta > 0$, define

$$Q_X(x^n) = \left[\frac{\Gamma(n+\delta)}{\Gamma(\delta)}\right]^{-1} \prod_{i=1}^{\alpha} \frac{\Gamma(n_{i,\cdot} + \delta/\alpha)}{\Gamma(\delta/\alpha)},$$

$$Q_Y(y^n) = \left[\frac{\Gamma(n+\delta)}{\Gamma(\delta)}\right]^{-1} \prod_{j=1}^{\beta} \frac{\Gamma(n_{\cdot,j} + \delta/\beta)}{\Gamma(\delta/\beta)},$$

$$Q_{XY}(x^n, y^n) = \left[\frac{\Gamma(n+\delta)}{\Gamma(\delta)}\right]^{-1} \prod_{i=1}^{\alpha} \prod_{j=1}^{\beta} \frac{\Gamma(n_{i,j} + \delta/\alpha\beta)}{\Gamma(\delta/\alpha\beta)}.$$

This is the BDeu (Bayesian Dirichlet equivalent uniform) scoring scheme [3], corresponding to prior frequencies $a_{i,j} = \delta/(\alpha\beta)$, $a_{i,\cdot} = \delta/\alpha$, $a_{\cdot,j} = \delta/\beta$. The same idea extends to three or more variables, and the conditional marginal likelihood is

$$Q_{X|Y}(x^n, y^n) := \frac{Q_{XY}(x^n, y^n)}{Q_Y(y^n)} = \prod_{j=1}^{\beta} \left\{ \left[\frac{\Gamma(n_{\cdot,j} + \delta/\beta)}{\Gamma(\delta/\beta)}\right]^{-1} \prod_{i=1}^{\alpha} \frac{\Gamma(n_{i,j} + \delta/\alpha\beta)}{\Gamma(\delta/\alpha\beta)} \right\}. \tag{8.9}$$

In general,

$$\frac{\prod_{i=1}^{\alpha}\prod_{j=1}^{\beta}\Gamma(n_{i,j}+a_{i,j})}{\Gamma(\sum_{i=1}^{\alpha}\sum_{j=1}^{\beta}(n_{i,j}+a_{i,j}))} \cdot \frac{\Gamma(\sum_{i=1}^{\alpha}\sum_{j=1}^{\beta}a_{i,j})}{\prod_{i=1}^{\alpha}\prod_{j=1}^{\beta}\Gamma(a_{i,j})} \cdot \frac{\Gamma(\sum_{j=1}^{\beta}(n_{\cdot,j}+a_{\cdot,j}))}{\prod_{j=1}^{\beta}\Gamma(n_{\cdot,j}+a_{\cdot,j})} \cdot \frac{\prod_{j=1}^{\beta}\Gamma(a_{\cdot,j})}{\Gamma(\sum_{j=1}^{\beta}a_{\cdot,j})}$$

holds. As with the earlier prior frequency schemes, BDeu assumes $a_{i,j}, a_{i,\cdot}, a_{\cdot,j}$ are constant across i, j. However, unlike Jeffreys' prior—where we took $a_{i,j} = a_{i,\cdot} = a_{\cdot,j} = a > 0$—BDeu uses $a_{i,j} = \delta/(\alpha\beta)$ and thus

$$a_{\cdot,j} = \sum_{i=1}^{\alpha} a_{i,j} = \frac{\delta}{\beta}, \qquad a_{i,\cdot} = \sum_{j=1}^{\beta} a_{i,j} = \frac{\delta}{\alpha}.$$

With $a = 1/2$ (Jeffreys' prior), we can write explicitly

$$Q_{X|Y}(x^n \mid y^n) = \prod_{j=1}^{\beta} \left\{ \frac{\Gamma(\alpha/2)}{\Gamma(n_{\cdot,j} + \alpha/2)} \prod_{i=1}^{\alpha} \frac{\Gamma(n_{i,j} + 1/2)}{\Gamma(1/2)} \right\}, \tag{8.10}$$

see Problem 89.

Although BDeu is more widely used than AIC, BIC, or Jeffreys' method, it suffers from a critical issue discussed below.

Let S be the parent set of X, and denote the empirical entropy and the number of parameters by $H(X \mid S)$ and $d(X, S)$, respectively. Define the information criteria

$$AIC(X, S) = H(X \mid S) + d(X, S), \qquad BIC(X, S) = H(X \mid S) + \frac{d(X, S)}{2} \log n.$$

Then

$$AIC(X, S) > AIC(X, S') \ (S \subset S') \Longrightarrow H(X \mid S) > H(X \mid S'), \tag{8.11}$$

$$BIC(X, S) > BIC(X, S') \ (S \subset S') \Longrightarrow H(X \mid S) > H(X \mid S'), \tag{8.12}$$

see Problem 90. We call this property the **regularity** of the information criterion.

If regularity fails—i.e., $H(X \mid S') > H(X \mid S)$ even though $S \subset S'$ (so S predicts X better), yet the score (AIC/BIC) prefers S'—then the model with better predictive ability is not selected, and an unnecessarily complex, redundant model is chosen. This inversion leads to serious problems:

- Unnecessary model complexity leads to **overfitting** and potentially spurious causal interpretations.
- During search, adding a parent need not correspond to an entropy (likelihood) improvement, making pruning difficult and complicating algorithm design [35].

Thus, for consistency and interpretability in structure learning, such regularity is desirable.

BDeu does not satisfy regularity [30].

Example 8.5 Let X, Y take α, β values, respectively, and consider $n \geq 2$ with $x^n = \underbrace{1, \ldots, 1}_{n}$, $y^n = \underbrace{1, \ldots, 1}_{n}$. Since X occurs independently of Y, we would like $Q_X(x^n) \geq Q_{X|Y}(x^n \mid y^n)$. Under Jeffreys' prior,

$$Q_X(x^n) = \frac{\Gamma(\alpha/2)}{\Gamma(n + \alpha/2)} \cdot \frac{\Gamma(n + 1/2)}{\Gamma(1/2)},$$

$$Q_Y(y^n) = \frac{\Gamma(\beta/2)}{\Gamma(n + \beta/2)} \cdot \frac{\Gamma(n + 1/2)}{\Gamma(1/2)},$$

$$Q_{XY}(x^n, y^n) = \frac{\Gamma(\alpha\beta/2)}{\Gamma(n + \alpha\beta/2)} \cdot \frac{\Gamma(n + 1/2)}{\Gamma(1/2)}, \tag{8.13}$$

$$Q_{X|Y}(x^n \mid y^n) = \left[\frac{\Gamma(\alpha\beta/2)}{\Gamma(n + \alpha\beta/2)}\right]\left[\frac{\Gamma(\beta/2)}{\Gamma(n + \beta/2)}\right]^{-1} \leq Q_X(x^n), \tag{8.14}$$

where (8.13) treats (X, Y) as a single variable with $\alpha\beta$ states. The inequality (8.14) follows because

$$a_n := \frac{\Gamma(n+\alpha\beta/2)\Gamma(n+1/2)}{\Gamma(n+\alpha/2)\Gamma(n+\beta/2)} \cdot \frac{\Gamma(\alpha/2)\Gamma(\beta/2)}{\Gamma(\alpha\beta/2)\Gamma(1/2)}, \quad n \geq 0$$

is nondecreasing with $a_0 = 1$ (Problem 91), and equality $a_{n+1} = a_n$ never holds for $n \neq 0$ (Problem 91). By contrast, under BDeu,

$$Q_X(x^n) = \frac{\Gamma(\delta)}{\Gamma(n+\delta)} \cdot \frac{\Gamma(n+\delta/\alpha)}{\Gamma(\delta/\alpha)}, \tag{8.15}$$

$$Q_Y(y^n) = \frac{\Gamma(\delta)}{\Gamma(n+\delta)} \cdot \frac{\Gamma(n+\delta/\beta)}{\Gamma(\delta/\beta)}, \tag{8.16}$$

$$Q_{XY}(x^n, y^n) = \frac{\Gamma(\delta)}{\Gamma(n+\delta)} \cdot \frac{\Gamma(n+\delta/(\alpha\beta))}{\Gamma(\delta/(\alpha\beta))},$$

$$Q_{X|Y}(x^n \mid y^n) = \left[\frac{\Gamma(n+\delta/(\alpha\beta))}{\Gamma(\delta/(\alpha\beta))}\right]\left[\frac{\Gamma(n+\delta/\beta)}{\Gamma(\delta/\beta)}\right]^{-1} \geq Q_X(x^n), \tag{8.17}$$

because

$$b_n := \frac{\Gamma(n+\delta)}{\Gamma(n+\delta/\alpha)} \frac{\Gamma(n+\delta/(\alpha\beta))}{\Gamma(n+\delta/\beta)} \cdot \frac{\Gamma(\delta/\alpha)\Gamma(\delta/\beta)}{\Gamma(\delta)\Gamma(\delta/(\alpha\beta))}, \quad n \geq 0$$

is nondecreasing with $b_0 = 1$ (Problem 91), and equality $b_{n+1} = b_n$ never holds for $n \neq 0$ (Problem 91). ■

Example 8.6 In the datasets (a) and (b) below, we have $H(X \mid Y) = H(X \mid Y, Z) = 0$: X is fully determined by Y, and adding Z does not reduce the conditional entropy.

From the viewpoint of regularity, we should select $\{Y\}$ as the parent set and avoid the empty set or the redundant sets $\{Z\}$ and $\{Y, Z\}$.

Under Jeffreys' prior,

$$Q_{X|Y}(x^n \mid y^n) > Q_{X|YZ}(x^n \mid y^n, z^n),$$

i.e., adding unnecessary Z lowers (worsens) the score, so Jeffreys' method preserves regularity. In contrast, under the BDeu score, extending (8.9) to $Q_{X|YZ}(x^n, y^n, z^n)$ via

$$\prod_{j=1}^{\beta}\prod_{k=1}^{\gamma}\left\{\left[\frac{\Gamma(n_{\cdot,j,k}+\delta/\beta\gamma)}{\Gamma(\delta/(\beta\gamma))}\right]^{-1}\prod_{i=1}^{\alpha}\frac{\Gamma(n_{i,j,k}+\delta/\alpha\beta\gamma)}{\Gamma(\delta/\alpha\beta\gamma)}\right\},$$

we obtain

$$Q_{X|Y}(x^n \mid y^n) < Q_{X|YZ}(x^n \mid y^n, z^n)$$

(Problem 93), meaning $\{Y, Z\}$—which should *not* be chosen—receives a higher score than $\{Y\}$. Hence, regularity is violated. This shows BDeu tends to include superfluous parents even when the data satisfy $H(X \mid Y) = 0$. A concrete example follows.

(a)

X	Y	Z
1	1	2
2	2	1
1	1	2
2	2	1

(b)

X	Y	Z
1	1	1
2	2	1
1	3	2
2	4	2
1	1	1
2	2	1
1	3	2
2	4	2

Here, (a) has $(\alpha, \beta, \gamma) = (2, 2, 2)$ and (b) has $(2, 4, 2)$. In (a), using $\Gamma(x+1) = x\Gamma(x)$ $(x > 0)$,

$$\frac{\Gamma(4/2)}{\Gamma(4+4/2)}\left[\frac{\Gamma(2/2)}{\Gamma(4+2/2)}\right]^{-1} = \frac{1}{5} > \frac{1}{7} = \frac{\Gamma(8/2)}{\Gamma(4+8/2)}\left[\frac{\Gamma(4/2)}{\Gamma(4+4/2)}\right]^{-1} \tag{8.18}$$

$$\left[\frac{\Gamma(4+1/2)}{\Gamma(1/2)}\right]^{-1}\frac{\Gamma(4+1/4)}{\Gamma(1/4)} = \frac{3\cdot 13}{2^4\cdot 7} < \frac{5\cdot 17}{2^4\cdot 13} = \left[\frac{\Gamma(4+1/4)}{\Gamma(1/4)}\right]^{-1}\frac{\Gamma(4+1/8)}{\Gamma(1/8)} \tag{8.19}$$

so, with Jeffreys' prior ($a = 1/2$) and with BDeu, the inequalities between $Q_{X|Y}(x^n, y^n)$ and $Q_{X|YZ}(x^n, y^n, z^n)$ are $>$ and $<$, respectively. The same can be confirmed for (b) (Problem 93). In BDeu, as in (b) when the number of states grows, this tendency to favor "extra" variables becomes even more pronounced. ■

8.6 Learning Forest Structures

So far, we have focused on constructing BNs from samples. With p variables, computation grows exponentially in p: Even if the ordering is known, finding $\pi_k \subseteq \{1, \ldots, k-1\}$ requires comparing 2^{k-1} candidate parent sets; hence $\sum_{k=1}^{p}(2^{k-1}) = 2^p - 1$ structures in total. If we also score over orderings, the time grows further.

In Chap. 3, the Chow–Liu algorithm approximates the joint distribution using the true mutual information $I(X_i, X_j)$. With only data, let X and Y take values in $\{1, \ldots, \alpha\}$ and $\{1, \ldots, \beta\}$, and denote counts by $n_{i,\cdot}$, $n_{\cdot,j}$, $n_{i,j}$ for $X = i$, $Y = j$, $(X, Y) = (i, j)$. A common estimator is

$$I_n := \sum_i \sum_j \frac{n_{i,j}}{n} \log \left(\frac{n_{i,j}}{n} \Big/ \frac{n_{i,\cdot}}{n} \frac{n_{\cdot,j}}{n} \right). \tag{8.20}$$

A Python implementation might be:

```
def I_n(x, y):
    x = np.asarray(x)
    y = np.asarray(y)
    if x.shape[0] != y.shape[0]:
        raise ValueError("Lengths of x and y do not match.")
    n = x.size

    # Relabel values to 0..K-1
    _, xi = np.unique(x, return_inverse=True)
    _, yj = np.unique(y, return_inverse=True)
    a = xi.max() + 1
    b = yj.max() + 1

    # Frequency table z[i,j], marginal frequencies u[i]=n_i., v[j]=n_.j
    z = np.zeros((a, b), dtype=int)
    np.add.at(z, (xi, yj), 1)
    u = z.sum(axis=1)    # shape (a,)
    v = z.sum(axis=0)    # shape (b,)

    # Mutual information
    S = 0.0
    ii, jj = np.nonzero(z)
    for i, j in zip(ii, jj):
        nij = z[i, j]
        S += (nij / n) * np.log((nij * n) / (u[i] * v[j]))
    return float(S)
```

Example 8.7 `Asia` is a dataset with $p = 8$ variables and $n = 5000$ samples (see Table 1.1).

Using `I_n` to estimate mutual information from the data, we built a weight matrix and ran `kruskal`. The resulting `edge_list` was visualized with `networkx` (Fig. 8.7).

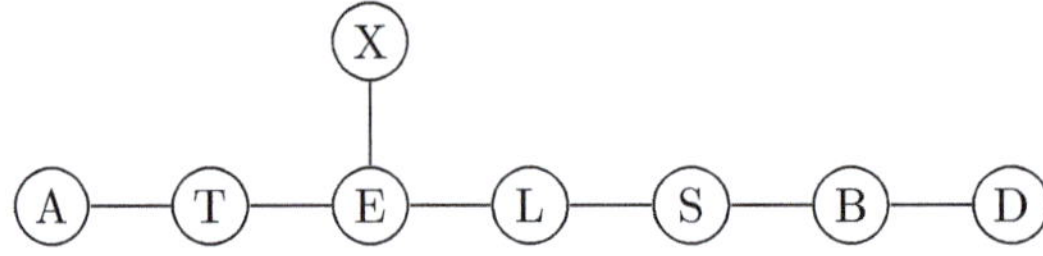

Fig. 8.7 Maximum spanning tree obtained by estimating mutual information for the Asia dataset using the function `I.n` (maximum likelihood method) and applying Kruskal's algorithm

```python
import numpy as np
import networkx as nx
import matplotlib.pyplot as plt
import bnlearn as bn

# Convert data to matrix
df = bn.import_example(data='asia')  # returns a pandas.DataFrame
x = df.to_numpy()
n, p = x.shape

# Weight matrix of mutual information
w = np.zeros((p, p), dtype=float)
for i in range(p - 1):
    xi = x[:, i]
    for j in range(i + 1, p):
        w[i, j] = I_n(xi, x[:, j])
        w[j, i] = w[i, j]

# Chow--Liu tree (maximum spanning tree)
edge_list = kruskal(w)   # assumes it returns a list of 0-based tuples (i, j)
print(edge_list)

# Visualization (networkx)
names = list(df.columns)
G = nx.Graph()
G.add_nodes_from(names)
for i, j in edge_list:
    G.add_edge(names[i], names[j], weight=float(w[i, j]))

pos = nx.spring_layout(G, seed=0)
nx.draw_networkx(G, pos, with_labels=True, arrows=False)
plt.title("Chow--Liu (I_n)")
plt.axis("off")
plt.show()
```

■

Here we extend Chow–Liu to learn a forest [27]. We still apply Kruskal's algorithm but focus on improving the estimator of mutual information.

So far, with X, Y taking values in $\{1, \ldots, \alpha\}$ and $\{1, \ldots, \beta\}$ and samples $(x_1, y_1), \ldots, (x_n, y_n)$, we have used (8.20) to estimate $I(X, Y)$. A benefit is that $I_n \to I(X, Y)$ in probability as $n \to \infty$.

However, even when $X \perp\!\!\!\perp Y$, for finite n we still get a positive value that only shrinks with n, so I_n is not directly usable for independence testing. We therefore consider another estimator, redefining from (8.6):

$$J_n := \max\left\{\frac{1}{n}\log\frac{Q_{XY}(x^n, y^n)}{Q_X(x^n)Q_Y(y^n)}, 0\right\}. \tag{8.21}$$

A Python version is:

```python
def _to_1based(a):
    """Relabel an arbitrary categorical array a to 1..K (and return K as well
      )."""
    _, inv = np.unique(a, return_inverse=True)
    return inv + 1, int(inv.max() + 1)
```

```
def J_n(x, y):
    x1, alpha = _to_1based(x)
    y1, beta  = _to_1based(y)
    z1 = (y1 - 1) * alpha + x1  # Cartesian-product labels 1..(alpha*beta)
    value = Q_4(x1, alpha) + Q_4(y1, beta) - Q_4(z1, alpha * beta)
    return max(float(value), 0.0)
```

The marginal likelihood function used here is:

```
def Q_4(x, alpha):
    x = np.asarray(x, dtype=int)
    n = x.size
    cc = np.zeros(alpha, dtype=int)  # count per category
    nc = 0
    logq = 0.0
    for v in x:
        k = v - 1  # 0-based
        logq -= math.log((cc[k] + 0.5) / (nc + alpha / 2.0))
        cc[k] += 1
        nc += 1
    return logq / n
```

For small n one can also use Q_1 from Chap. 7, but factorials require accumulating in log form as above.

By Proposition 8.2, we have

$$J_n = 0 \iff X \perp\!\!\!\perp Y$$

with probability 1. Moreover,

$$J_n = \max\left\{I_n - \frac{1}{2n}(\alpha - 1)(\beta - 1)\log n + O\left(\frac{1}{n}\right),\ 0\right\} \tag{8.22}$$

(Problem 96). Here the second term penalizes model complexity (more states), which can be interpreted as controlling **overfitting**.

The forest-learning algorithm I proposed in 1993 [27] effectively used (8.22) in place of (8.21) within Chow–Liu (see also [29]). In Python:

```
def J_n(x, y):
    x = np.asarray(x); y = np.asarray(y)
    n = x.size
    alpha = np.unique(x).size
    beta  = np.unique(y).size
    val = I_n(x, y) - ((alpha - 1) * (beta - 1)) * math.log(n) / (2.0 * n)
    return max(float(val), 0.0)
```

The learned BN is a forest (not a single tree) because disconnected independent components have no path between them. Also, when using J_n, the order in which edges are added generally differs from that under I_n; the second term in J_n mitigates overfitting, with larger $(\alpha - 1)(\beta - 1)$ implying a stronger tendency to overfit.

If we replace I_n by J_n in Example 8.7 (Fig. 8.7), we obtain Fig. 8.8. The node *Asia* is separated from the others, indicating low mutual information with the rest

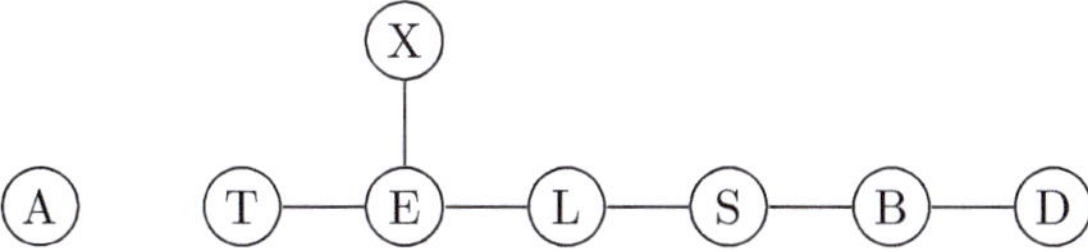

Fig. 8.8 Maximum spanning tree obtained by estimating mutual information with function `J.n` (Bayesian method) for the Asia dataset and applying Kruskal's algorithm. Unlike Fig. 8.7, Asia is separated

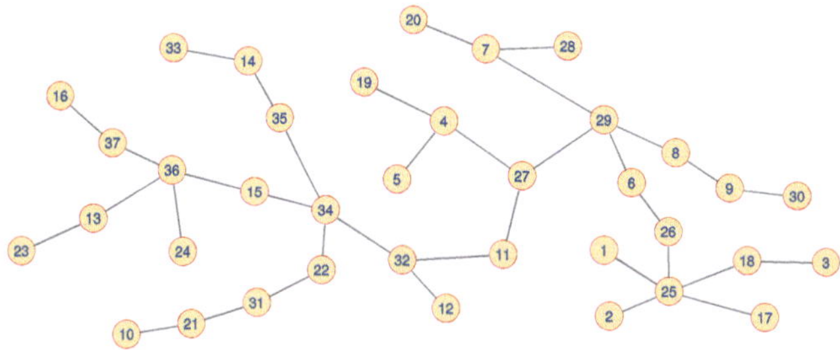

Fig. 8.9 Chow–Liu applied to the Alarm dataset with $p = 37$. In this example, the forests obtained using I_n and J_n (from (8.21)) coincide

and selection of independence. Using the J_n in (8.22) gives a similar result. Since all variables here are binary, $(\alpha - 1)(\beta - 1)$ is constant, so the edge-joining order is the same (Fig. 8.9).

From Propositions 8.3 and 8.4, in the Gaussian case we can define

$$I_n := -\frac{1}{2}\log(1 - \hat{\rho}_n^2), \qquad J_n := -\frac{1}{2}\log(1 - \hat{\rho}_n^2) - \frac{1}{2n}\log n.$$

In Python:

```
import numpy as np, math
```

```
def I_n_gauss(x, y):
    """I_n␣=␣-1/2␣*␣log(1␣-␣rho^2)"""
    x = np.asarray(x, dtype=float)
    y = np.asarray(y, dtype=float)
    rho = np.corrcoef(x, y)[0, 1]           # correlation equivalent to R's
        cov/var
    rho = float(np.clip(rho, -1.0, 1.0))    # numerical stabilization
    return -0.5 * math.log(max(1.0 - rho * rho, 1e-300))
```

```
def J_n_gauss(x, y):
    """J_n␣=␣I_n␣-␣(1/(2n))␣*␣log␣n,␣lower-clipped␣at␣0"""
    n = len(x)
    val = I_n_gauss(x, y) - 0.5 * math.log(n) / n
    return max(val, 0.0)
```

As in the discrete case, $J_n \to I(X, Y)$ as $n \to \infty$, and $J_n = 0$ if and only if $X \perp\!\!\!\perp Y$ (Problem 98).

Appendix

The proof of Proposition 8.1 follows Theorem 2 of [37].

Proof of Proposition 8.1

For any pairwise disjoint $S, T, U \subseteq \{1, \ldots, p\}$ $(S, T \neq \{\})$,

$$S \perp\!\!\!\perp_G T \mid U \Longrightarrow X_S \perp\!\!\!\perp X_T \mid X_U \tag{8.23}$$

must hold, and the relation must fail if any edge in E is removed (minimality). Such a G is called a BN (Chap. 3).

We prove by induction that for each $2 \leq k \leq p$, the graph with $V = \{1, \ldots, k\}$ and $E = \cup_{j=1}^{k}\{(i, j) \mid i \in \pi_j\}$ is a BN for $X_1, \ldots, X_k$. For $k = 2$, the only nonempty disjoint pairs S, T with $U = \{\}$ are $\{1\}, \{2\}$ and $\{2\}, \{1\}$, so (8.23) holds. Also $E = \{\}$, so minimality holds.

Assume $V = \{1, \ldots, k-1\}$ and $E = \cup_{j=1}^{k-1}\{(i, j) \mid i \in \pi_j\}$ define a BN for $X_1, \ldots, X_{k-1}$. After adding vertex k and edges (j, k), $j \in \pi_k$, we show

$$S \perp\!\!\!\perp_G T \mid U \Longrightarrow X_S \perp\!\!\!\perp X_T \mid X_U, \tag{8.24}$$

$$(S \cup k) \perp\!\!\!\perp_G T \mid U \Longrightarrow X_{S\cup k} \perp\!\!\!\perp X_T \mid X_U, \tag{8.25}$$

$$S \perp\!\!\!\perp_G T \mid (U \cup k) \Longrightarrow X_S \perp\!\!\!\perp X_T \mid X_{U\cup k}. \tag{8.26}$$

(We abbreviate $S \cup \{k\}$ by $S \cup k$.) Establishing (8.24) for $V = \{1, \ldots, k\}$ and $E = \cup_{j=1}^{k}\{(i, j) \mid i \in \pi_j\}$ suffices.

For (8.24), only edges into k are added, creating colliders at k and leaving paths between S and T unchanged; X_k does not affect conditional independences among $X_1, \ldots, X_{k-1}$, so (8.24) holds.

For (8.25), note by construction

$$X_k \perp\!\!\!\perp X_{\overline{\pi}_k} \mid X_{\pi_k}, \tag{8.27}$$

where $\overline{A}$ denotes the complement of A in $\{1, \ldots, k-1\}$. Let

$$\pi_* := \pi_k \cap \overline{S \cup T \cup U}, \qquad \pi^* := \overline{\pi}_k \cap \overline{S \cup T \cup U} = \overline{\pi_k \cup S \cup T \cup U}$$

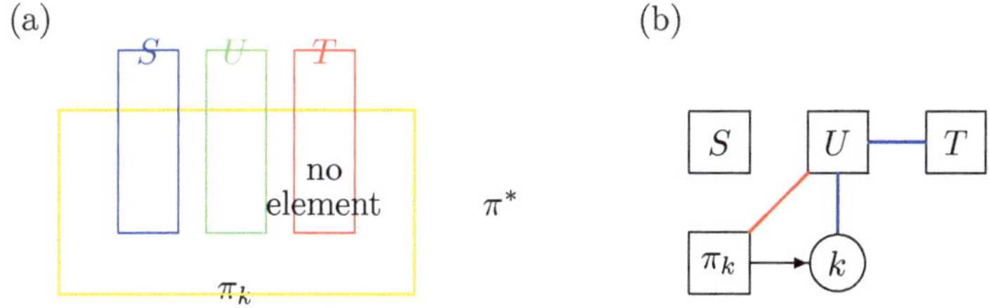

Fig. 8.10 Proof of Proposition 8.1. (**a**) Since $T \cap \pi_k = \{\}$, it follows that $\overline{T \cup \pi^*} = S \cup U \cup \pi_k$. (**b**) If $k \perp\!\!\!\perp_G T \mid U$, then it must also hold that $\pi_k \perp\!\!\!\perp_G T \mid U$; otherwise, there exists a path from k to T through π_k

(Fig. 8.10a). Apply (3.6) with

$$A := k, \quad B := (T \cup \pi^*) \cap \overline{\pi}_k, \quad C := \pi_k, \quad D := \overline{T \cup \pi^*} \cap \overline{\pi}_k,$$

to obtain from (8.27)

$$X_k \perp\!\!\!\perp X_{(T \cup \pi^*) \cap \overline{\pi}_k} \mid X_{\pi_k \cup \overline{T \cup \pi^*} \cap \overline{\pi}_k}.$$

Since $T \cap \pi_k = \{\}$ (otherwise an edge $j \to k$ with $j \in T$ yields a length-1 directed path contradicting $(S \cup k) \perp\!\!\!\perp_G T \mid U$ unless U blocks it), we have

$$\begin{aligned} (T \cup \pi^*) \cap \overline{\pi}_k &= T \cup (\pi^* \cap \pi_k) = T \cup \pi^*, \\ \overline{T \cup \pi^*} &= \overline{T} \cap \overline{\pi^*} = \overline{T} \cap (\pi_k \cup S \cup T \cup U) = S \cup U \cup \pi_k, \\ \pi_k \cup \overline{T \cup \pi^*} \cap \overline{\pi}_k &= \pi_k \cup \overline{T \cup \pi^*} = S \cup U \cup \pi_k, \end{aligned}$$

hence

$$X_k \perp\!\!\!\perp X_{T \cup \pi^*} \mid X_{S \cup U \cup \pi_k}.$$

Applying (3.5) yields

$$X_k \perp\!\!\!\perp X_T \mid X_{S \cup U \cup \pi_k}. \tag{8.28}$$

From $k \perp\!\!\!\perp_G T \mid U$ (which must hold, or else some $t \in T$ is not d-separated from k) and $S \perp\!\!\!\perp_G T \mid U$, we get $(S \cup \pi_k) \perp\!\!\!\perp_G T \mid U$, which by (8.24) implies

$$X_{S \cup \pi_k} \perp\!\!\!\perp X_T \mid X_U. \tag{8.29}$$

Apply (3.7) with

$$A := T, \quad B := S \cup \pi_k, \quad C := U, \quad D := k$$

to obtain

$$X_{S \cup k \cup \pi_k} \perp\!\!\!\perp X_T \mid X_U,$$

and then (3.5) gives the RHS of (8.25).

For (8.26), assume $S \perp\!\!\!\perp_G T \mid (U \cup k)$. Any path between $s \in S$ and $t \in T$ must hit k as a collider and thus is not blocked by k. Therefore either $k \perp\!\!\!\perp_G S \mid U$ or $k \perp\!\!\!\perp_G T \mid U$ holds. WLOG assume $k \perp\!\!\!\perp_G T \mid U$. Then by (3.8),

$$S \perp\!\!\!\perp_G T \mid U \text{ and } k \perp\!\!\!\perp_G T \mid U \Longleftrightarrow (S \cup k) \perp\!\!\!\perp_G T \mid U.$$

Applying (8.25) yields $X_{S \cup k} \perp\!\!\!\perp X_T \mid X_U$, and finally (3.6) gives $X_S \perp\!\!\!\perp X_T \mid X_{U \cup k}$, i.e., (8.26).

Thus (8.24) holds for disjoint $S, T, U \subseteq \{1, \ldots, k\}$. Moreover, by construction we have a boundary DAG (Sect. 8.1), so removing any edge violates (8.24); minimality holds and the proposition follows. ■

Proof of Proposition 8.3

Let

$$Q_X = \pi^{-n/2} \frac{\Gamma(\nu_n/2)}{\Gamma(\nu_0/2)} \frac{\lambda_{X,0}^{\nu_0/2}}{\lambda_{X,n}^{\nu_n/2}} \left(\frac{\kappa_0}{\kappa_n}\right)^{1/2}, \quad Q_Y = \pi^{-n/2} \frac{\Gamma(\nu_n/2)}{\Gamma(\nu_0/2)} \frac{\lambda_{Y,0}^{\nu_0/2}}{\lambda_{Y,n}^{\nu_n/2}} \left(\frac{\kappa_0}{\kappa_n}\right)^{1/2},$$

$$Q_{XY} = \pi^{-n} \frac{\Gamma_2(\nu_n/2)}{\Gamma_2(\nu_0/2)} \frac{\det(\Lambda_0)^{\nu_0/2}}{\det(\Lambda_n)^{\nu_n/2}} \left(\frac{\kappa_0}{\kappa_n}\right)^{1},$$

$$\lambda_{X,n} = \lambda_{X,0} + (\bar{x} - \mu_{X,0})^2 + \sum_{i=1}^{n} (x_i - \bar{x})^2, \quad \lambda_{Y,n} = \lambda_{Y,0} + (\bar{y} - \mu_{Y,0})^2 + \sum_{i=1}^{n} (y_i - \bar{y})^2,$$

$$\lambda_n = \lambda_0 + (\bar{x} - \mu_{X,0})(\bar{y} - \mu_{Y,0}) + \sum_{i=1}^{n} (x_i - \bar{x})(y_i - \bar{y}),$$

$$\Lambda_0 = \begin{bmatrix} \lambda_{X,0} & \lambda_0 \\ \lambda_0 & \lambda_{Y,0} \end{bmatrix}, \quad \Lambda_n = \begin{bmatrix} \lambda_{X,n} & \lambda_n \\ \lambda_n & \lambda_{Y,n} \end{bmatrix}.$$

Since

$$\hat{\rho}_n^2 - \frac{\lambda_n^2}{\lambda_{X,n}\lambda_{Y,n}} \to 0 \text{ in probability},$$

it suffices to show

$$-\log \frac{\Gamma_2(\nu_n/2)}{\Gamma_2(\nu_0/2)} = \frac{1}{2}\log n + O(1). \tag{8.30}$$

Note

$$\frac{\Gamma_2(\nu_n/2)}{\Gamma(\nu_n/2)} = \pi^{1/2}\frac{\Gamma(\frac{\nu_n}{2})\Gamma(\frac{\nu_n-1}{2})}{\Gamma(\frac{\nu_n}{2})^2} = \pi^{1/2}\frac{\Gamma(\frac{\nu_n-1}{2})}{\Gamma(\frac{\nu_n}{2})}.$$

Since $\log\Gamma(x)$ is convex, Gautschi's inequality gives for $(u, v) = (\frac{\nu_n-2}{2}, \frac{\nu_n}{2})$ and $(\frac{\nu_n-1}{2}, \frac{\nu_n+1}{2})$

$$\Gamma\left(\frac{\nu_n-1}{2}\right) < \left(\frac{\nu_n-2}{2}\right)^{-1/2}\Gamma\left(\frac{\nu_n}{2}\right), \qquad \Gamma\left(\frac{\nu_n}{2}\right) < \left(\frac{\nu_n-1}{2}\right)^{1/2}\Gamma\left(\frac{\nu_n-1}{2}\right).$$

Hence

$$\sqrt{\frac{\nu_n-2}{2}} < \frac{\Gamma(\frac{\nu_n}{2})}{\Gamma(\frac{\nu_n-1}{2})} < \sqrt{\frac{\nu_n-1}{2}},$$

and therefore

$$\frac{1}{2}\log(\nu_n-2) - \frac{1}{2}\log 2\pi < \log\frac{\Gamma(\frac{\nu_n}{2})^2}{\Gamma_2(\frac{\nu_n}{2})} < \frac{1}{2}\log(\nu_n-1) - \frac{1}{2}\log 2\pi.$$

Thus (8.30) holds. ∎

Problems 79–100

79. As in Sect. 7.1 where the BN structure for the `alarm` dataset was obtained by hill climbing (BIC), compute `edges` for the `asia` dataset as well, and visualize the structure using `igraph`.

```
# pip install pandas networkx matplotlib pgmpy
import numpy as np
import pandas as pd
import networkx as nx
import matplotlib.pyplot as plt
from pgmpy.estimators import HillClimbSearch, BicScore
```

```
# --- Load data (8 columns: A, S, T, L, B, E, X, D) ---
df = pd.read_csv("asia.csv").astype("category")  # e.g., bnlearn's asia
    saved as CSV
# ========== BIC ==========
bic_model = HillClimbSearch(df).estimate(scoring_method=BicScore(df))
edges_bic = list(bic_model.edges())
print("BIC edges:", edges_bic)

# Visualization (networkx)
G = nx.DiGraph()
G.add_nodes_from(["A","S","T","L","B","E","X","D"])
G.add_edges_from(edges_bic)
pos = nx.spring_layout(G, seed=0)
nx.draw(G, pos, with_labels=True, node_color="yellow", arrows=True)
plt.title("Asia (HillClimb + BIC)")
plt.show()
```

Also, replace `BIC` with `AIC` and run it. Compare the structures—what differences do you observe?

80. In Example 8.1, how does the function `ftable_py` differ from ordinary table operations that compute row-wise and column-wise sums?
81. For two pairs (H, d) and (H', d') of empirical entropy and parameter count corresponding to different parent sets, show that

$$H + \frac{d}{2}\log n \le H' + \frac{d'}{2}\log n, \qquad H + d \ge H' + d'$$

can hold simultaneously only when $d \le d'$. In other words, BIC imposes a larger penalty for complexity than AIC and therefore prefers simpler structures.

82. The function `IC_min` in Sect. 7.2 is implemented for continuous variables. Rewrite it for discrete variables, and apply it to the `asia` dataset; compare the resulting structure with that obtained by the hill climbing method in Sect. 7.1.
83. The following function `Q` takes an $n \times p$ matrix as input and returns the **negative mean of the log marginal likelihood** over the p variables (implemented by sequentially adding Jeffreys' prior 1/2). The example computes only 3 cases among the scores corresponding to the 11 Markov equivalence classes for 3 variables. Evaluate all of them, and find the BN that **maximizes the marginal likelihood (i.e., minimizes the negative log mean)**.

```
import numpy as np, math
```

```
def Q(x):
    x = np.asarray(x)
    if x.ndim == 1:
        x = x.reshape(-1, 1)
    n, p = x.shape
    m = np.array([len(np.unique(x[:, j])) for j in range(p)], dtype=int
        )
    M = int(np.prod(m))
    cc = np.zeros(M, dtype=int)
```

```
    q = 0.0
    for i in range(n):  # i = 0..n-1
        z = int(x[i, 0] - 1)
        for j in range(1, p):
            z = z * m[j] + int(x[i, j] - 1)
        q -= np.log((cc[z] + 0.5) / (i + 0.5 * M))
        cc[z] += 1
    return q / n

# Data generation
n = 1000
u = np.random.binomial(1, 0.5, size=n)
v = (u + np.random.binomial(1, 0.1, size=n)) % 2
w = (v + np.random.binomial(1, 0.1, size=n)) % 2
x = u + 1
y = v + 1
z = w + 1

# Example runs
print(Q(x) + Q(y) + Q(z))
print(Q(x) + Q(np.column_stack([y, z])))
print(Q(np.column_stack([x, y])) + Q(np.column_stack([y, z])) - Q(np.
   column_stack([y, z])) - Q(y))
```

Hint: Smaller values (i.e., larger marginal likelihoods) are preferable.

84. In the proof of Proposition 8.2, prove that when X, Y are independent, $K_n = 0$ holds regardless of the values of k_X, k_Y, k_{XY}. Further, show that (8.5) holds in this case. *Hint:* Use

$$\alpha\beta - 1 - (\alpha - 1) - (\beta - 1) = (\alpha - 1)(\beta - 1).$$

85. Explain the positions of the checkmarks (✓) from the second row onward in Table 8.1.
86. Perform the same kind of operations as in (8.8) for the MN in Fig. 3.7. How can the eight scores be computed?
87. For BN structure learning without assuming an ordering:

    a. The first half of the procedure can be written as follows when the data matrix **x** ($n \times p$) and p are given. It is a program to find the minimum for AIC. Line 10 checks whether $x \notin S$. The loop from line 12 compares the values $R(X \mid S \setminus \{Y\})$ while removing each $Y \in S$ one at a time. After that loop, it compares with $Q(X \mid S)$. There is a double loop over `m` and `i`. The parent set is stored in y[m,i]\$, and its score in \verbz[m,i]@.

```
import numpy as np, math
```

```
empty = np.zeros((n, 0))
L = 2**p
y = -np.ones((p, L), dtype=int)
z = -np.ones((p, L))
for m in range(p):
    y[m, 0] = 1
    z[m, 0] = AIC_value(empty, x[:, m])
    for i in range(1, L):
```

```
            r = decimal_to_binary(i - 1, p)
            if r[m] == 0:
                z[m, i] = math.inf
                for j in range(p):
                    if r[j] == 1:
                        s = r.copy()
                        s[j] = 0
                        k = binary_to_decimal(s) + 1
                        if z[m, k] < z[m, i]:
                            z[m, i] = z[m, k]
                            y[m, i] = y[m, k]
                set_idx = decimal_to_set(i - 1)
                if set_idx:
                    aic = AIC_value(x[:, set_idx], x[:, m])
                else:
                    aic = AIC_value(empty, x[:, m])
                if aic < z[m, i]:
                    z[m, i] = aic
                    y[m, i] = i
```

After understanding the above program, run:

```
n = 500    # sample size
p = 4      # number of variables
x = np.zeros((n, p))
x[:, 0] = np.random.randn(n)                          # X1: standard normal
x[:, 1] = np.random.randn(n)                          # X2: standard normal
x[:, 2] = x[:, 0] + x[:, 1] + np.random.randn(n)   # X3 depends on X1
    , X2
x[:, 3] = x[:, 2] + np.random.randn(n)              # X4 depends on X3
```

Assume the other functions are defined as follows:

```
def decimal_to_binary(x, p):
    if x == 0:
        return []
    else:
        return [x % 2] + decimal_to_binary(x // 2, p) + [0] * p
```

```
def binary_to_decimal(x):
    if len(x) == 0:
        return 0
    else:
        return x[0] + 2 * binary_to_decimal(x[1:])
```

b. For the second half of the procedure, the loop starting at line 12 compares the values $T(S \setminus \{Y\})\, R(X \mid S \setminus \{Y\})$ while removing each $Y \in S$ one at a time. After that loop, it compares with $Q(X \mid S)$. There is a double loop over `m` and `i`. The score is stored in `u[i]`, the parent set in `v[i]`, and the set of variables after removal in `w[i]`. On line 30 the removed variables are obtained.

Up to line 22, the optimal order and parent sets have been obtained. Variable `t` at line 27 contains multiple (m) variables; draw directed edges from each of them toward `j`. Line 31 appends a row representing an edge.

The operation `k = w[k]` at line 32 obtains the set of variables one level upstream.

```
empty = np.zeros((n, 0))
L = 2**p
u = np.empty(L)
v = np.empty(L, dtype=int)
w = np.empty(L, dtype=int)
v[0] = 1
u[0] = 0
w[0] = 1

for i in range(1, L):
    r = decimal_to_binary(i - 1, p)
    u[i] = np.inf
    for j in range(p):
        if r[j] == 1:
            s = r.copy()
            s[j] = 0
            k = binary_to_decimal(s) + 1
            if u[k - 1] + z[j, k - 1] < u[i]:
                u[i] = u[k - 1] + z[j, k - 1]
                v[i] = y[j, k - 1]
                w[i] = k

edges = np.empty((0, 2), dtype=int)
k = L
r = decimal_to_set(k - 1)
while w[k - 1] > 1:
    t = decimal_to_set(v[k - 1] - 1)
    m = len(t)
    s = decimal_to_set(w[k - 1] - 1)
    j = list(set(r) - set(s))
    if m > 0 and len(j) > 0:
        edges = np.vstack([edges, np.column_stack([t, [j[0]] * m])])
    k = w[k - 1]
    r = s

edges
```

After understanding the above program, run:

```
import networkx as nx
import matplotlib.pyplot as plt

G = nx.DiGraph()
G.add_edges_from(edges.tolist())
labels = {i: str(i+1) for i in range(p)}
pos = nx.spring_layout(G, seed=0)
nx.draw(G, pos, labels=labels, with_labels=True, arrows=True)
plt.show()
```

88. For BN structure learning without assuming an ordering, when there are three variables X, Y, Z, enumerate all expressions for $R(\cdot \mid \cdot)$ and $T(\cdot)$. *Hint:* There are 12 and 7 expressions, respectively.
89. Prove (8.10).
90. Prove (8.11) and (8.12).
91. Show that the sequences $\{a_n\}$ and $\{b_n\}$ are monotonically nonincreasing. Also show that for $n \geq 1$, the equalities in (8.14) and (8.17) do not hold.

92. Replace the function Q in Problem 83 with the BDeu function for $\delta = 1$ and run it.
93. In Example 8.6, verify the inequalities (8.18) and (8.19) for cases (a) and (b). Explain why each inequality corresponds to the relationship between $Q_{X|Y}$ and $Q_{X|YZ}$ under Jeffreys' prior and under BDeu ($\delta = 1$), respectively. Perform the analogous check for (b) as well.
94. Suppose we observe n independent trials of (X, Y), and the marginal and joint frequencies are $c_X(x)$, $c_Y(y)$, and $c_{XY}(x, y)$ for $X = x \in A$, $Y = y \in B$. We estimate the mutual information by

$$I_n(x^n, y^n) := \sum_{x\in A}\sum_{y\in B} \frac{c_{XY}(x, y)}{n} \log \frac{\frac{c_{XY}(x, y)}{n}}{\frac{c_X(x)}{n} \cdot \frac{c_Y(y)}{n}}.$$

In the code below, `multi` generates random samples, and `I_n` computes the mutual information estimate. Fill in the blanks and run the code.

```
import numpy as np
import matplotlib.pyplot as plt

# Generate n draws from a multinomial with category probs 'prob'
def multi(n, prob):
    prob = np.asarray(prob, dtype=float)
    if not np.isclose(prob.sum(), 1.0):
        raise ValueError("prob␣must␣sum␣to␣1.")
    u = np.random.rand(n)
    y = np.empty(n, dtype=int)
    m = len(prob)
    for i in range(n):
        t = u[i]
        for j in range(m):
            if t < prob[j]:
                y[i] = j + 1            # 1-based categories
                break
            # (fill here) subtract prob[j] and continue
            t -= prob[j]
    return y

# ---- Experiment ----
r = 100   # repetitions

# Independent X, Y
S = []
for _ in range(r):
    x = multi(100, [1/2, 1/4, 1/4])
    y = multi(100, [1/2, 1/2])
    S.append(I_n(x, y))

# Dependent X, Y
T = []
for _ in range(r):
    x = multi(100, [1/2, 1/4, 1/4])
    z = multi(100, [8/10, 1/10, 1/10])
    y = ((x + z - 1) % 2) + 1
    T.append(I_n(x, y))

S = np.array(S)
T = np.array(T)
```

```
# ---- Visualization (boxplots) ----
plt.boxplot([S, T], labels=["Independent", "Dependent"])
plt.title("Estimated␣mutual␣information")
plt.ylabel("I_n")
plt.show()
```

95. Using the Asia dataset, generate a forest or tree with the functions `I_n`, `J_n`, and `kruskal`.
96. Prove (8.22). Also verify on both the `alarm` and `asia` datasets that the Bayesian Chow–Liu algorithm using (8.21) and the one using (8.22) construct the same forest.
97. For the 11 continuous variables (columns 1, 3, 5, 6, 7, 8, 10, 11, 12, 13, 14) in the Boston dataset, construct a forest using the method at the end of Chap. 7 (with both `I_n` and `J_n`).
98. For J_n defined in the Gaussian case, as in the discrete case one can define a function that converges to the true mutual information $I(X, Y)$ as $n \to \infty$ and equals 0 if and only if X and Y are independent. Below is the function corresponding to `I_n` for the discrete case. Similarly construct the function corresponding to `J_n`, and with $n = 100$ generate random samples for the independent and dependent cases to examine whether independence can be detected.

```
def I_n(x, y):
    # Build contingency table
    tab = np.zeros((len(np.unique(x)), len(np.unique(y))), dtype=int)
    x_vals = np.unique(x)
    y_vals = np.unique(y)
    for i, xv in enumerate(x_vals):
        for j, yv in enumerate(y_vals):
            tab[i, j] = np.sum((x == xv) & (y == yv))

    n = np.sum(tab)
    px = tab.sum(axis=1) / n
    py = tab.sum(axis=0) / n
    pxy = tab / n

    s = 0.0
    for i in range(tab.shape[0]):
        for j in range(tab.shape[1]):
            if pxy[i, j] > 0:
                s += pxy[i, j] * np.log(pxy[i, j] / (px[i] * py[j]))
    return s

# Example check
np.random.seed(0)
x = np.random.choice([1, 2, 3], size=100, p=[0.5, 0.25, 0.25])
y = np.random.choice([1, 2], size=100, p=[0.5, 0.5])
print(I_n(x, y))
```

99. Explain the steps from (8.27) to (8.28).
100. When constructing a forest using J_n, the order in which edges are added can differ from that using I_n. Show that when all random variables are binary, the edge-joining order is the same for both. Why is that?

Bibliography

1. T. W. Anderson. *An Introduction to Multivariate Statistical Analysis*. Wiley Series in Probability and Statistics. Wiley-Interscience, Hoboken, NJ, 3rd edition, 2003.
2. Salomon Bochner. *Lectures on Fourier Integrals*. Princeton University Press, Princeton, NJ, 1959. Originally published in 1932.
3. Wray Buntine. Theory refinement on Bayesian networks. In *Proceedings of the Seventh Conference on Uncertainty in Artificial Intelligence (UAI)*, pages 52–60. Morgan Kaufmann, 1991.
4. C. K. Chow and C. N. Liu. Approximating discrete probability distributions with dependence trees. *IEEE Transactions on Information Theory*, 14(3):462–467, 1968.
5. Georges Darmois. Analyse générale des liaisons stochastiques: étude particulière de l' analyse factorielle linéaire. *Revue de l'Institut International de Statistique*, 21(1/3):2–8, 1953.
6. J. J. Daudin. Partial association measures and an application to qualitative regression. *Biometrika*, 67(3):581–590, 1980.
7. Jayanta K. Ghosh, Mohan Delampady, and Tapas Samanta. *An Introduction to Bayesian Analysis: Theory and Methods*. Springer Texts in Statistics. Springer, New York, 2006.
8. Arthur Gretton, Kenji Fukumizu, Choon H. Teo, Le Song, Bernhard Schölkopf, and Alex J. Smola. A kernel statistical test of independence. In *Advances in Neural Information Processing Systems 20 (NeurIPS 2007)*, pages 585–592. Curran Associates, Inc., 2008.
9. Aapo Hyvärinen and Erkki Oja. Independent component analysis: algorithms and applications. *Neural networks*, 13(4–5):411–430, 2000.
10. Yutaka Kano and Shohei Shimizu. Causal inference using nonnormality. In *Proceedings of the International Symposium on Science of Modeling: The 30th Anniversary of the Information Criterion*, Tokyo, Japan, 2003.
11. Donald Kelker. *Normal Distribution: Characterizations with Applications*, volume 100 of *Lecture Notes in Statistics*. Springer, 1995. Revised October 29, 2008.
12. Joseph B. Kruskal. On the shortest spanning subtree of a graph and the traveling salesman problem. *Proceedings of the American Mathematical Society*, 7(1):48–50, 1956.
13. Steffen L. Lauritzen. *Graphical Models*. Oxford University Press, Oxford, 1996.
14. Steffen L. Lauritzen and David J. Spiegelhalter. Local computations with probabilities on graphical structures and their application to expert systems (with discussion). *Journal of the Royal Statistical Society: Series B (Statistical Methodology)*, 50(2):157–224, 1988.
15. Yuri V. Linnik. *Decomposition of Probability Distributions*. Oliver and Boyd, Edinburgh, 1968. Translated from the Russian edition (1962).
16. J. Marcinkiewicz. Sur une propriété de la loi de gauss. *Mathematicae*, 9:193–196, 1939.

J. Suzuki, *Graphical Models and Causal Discovery with Python*,
https://doi.org/10.1007/978-981-95-5308-2

17. Robb J. Muirhead. *Aspects of Multivariate Statistical Theory*. Wiley Series in Probability and Mathematical Statistics. Wiley, 1982.
18. Kevin P. Murphy. *Machine Learning: A Probabilistic Perspective*. Adaptive Computation and Machine Learning series. MIT Press, Cambridge, MA, 2012.
19. Judea Pearl. *Probabilistic Reasoning in Intelligent Systems: Networks of Plausible Inference*. Morgan Kaufmann, San Mateo, CA, 1988.
20. Robert C. Prim. Shortest connection networks and some generalizations. *Bell System Technical Journal*, 36(6):1389–1401, 1957.
21. Shohei Shimizu, Patrik O. Hoyer, Aapo Hyvärinen, and Auli Kerminen. A linear non-Gaussian acyclic model for causal discovery. *Journal of Machine Learning Research*, 7:2003–2030, 2006.
22. Shohei Shimizu, Takanori Inazumi, Yuichiro Sogawa, Aapo Hyvärinen, Yoshinobu Kawahara, Takashi Washio, Patrik O. Hoyer, and Kenneth A. Bollen. DirectLiNGAM: A direct method for learning a linear non-Gaussian structural equation model. *Journal of Machine Learning Research*, 12:1225–1248, 2011.
23. Tomi Silander and Petri Myllymäki. A simple approach for finding the globally optimal Bayesian network structure. In *Proceedings of the 28th Conference on Uncertainty in Artificial Intelligence (UAI)*, pages 476–485, 2012.
24. Ajit Singh and Andrew W. Moore. Finding optimal Bayesian networks by dynamic programming. Technical Report CMU-CALD-05-106, Center for Automated Learning and Discovery, Carnegie Mellon University, 2005.
25. V. P. Skitovich. On a property of the normal distribution. *Doklady Akademii Nauk SSSR*, 89:217–219, 1953.
26. Peter Spirtes, Clark Glymour, and Richard Scheines. Causation, prediction, and search. *MIT Press*, 2000. 2nd edition.
27. Joe Suzuki. Learning Bayesian belief networks based on the mdl principle: An efficient algorithm using branch and bound technique. In *Proceedings of the Ninth Conference on Uncertainty in Artificial Intelligence*, pages 266–273, Washington, D.C., 1993. Morgan Kaufmann.
28. Joe Suzuki. *Introduction to Bayesian Networks: Fundamentals of Probabilistic Knowledge Processing*. Baifukan, Tokyo, 2009.
29. Joe Suzuki. The Bayesian Chow-Liu algorithm. In *Proceedings of the 6th European Workshop on Probabilistic Graphical Models (PGM 2012)*, pages 315–322, Granada, Spain, 2012.
30. Joe Suzuki. A theoretical analysis of the BDeu scores in Bayesian network structure learning. *Behaviormetrika*, 44(1):97–116, 2017.
31. Joe Suzuki. *Statistical Learning with Math and R: 100 Exercises for Building Logic*. Springer, Singapore, 2020.
32. Joe Suzuki. *Statistical Learning with Math and Python: 100 Exercises for Building Logic*. Springer, Singapore, 2021.
33. Joe Suzuki. *Kernel Methods for Machine Learning with Math and Python: 100 Exercises for Building Logic*. Springer, Singapore, 2022.
34. Joe Suzuki. *Kernel Methods for Machine Learning with Math and R: 100 Exercises for Building Logic*. Springer, Singapore, 2022.
35. Joe Suzuki and Jun Kawahara. Branch and bound for regular Bayesian network structure learning. In *Proceedings of the 33rd Conference on Uncertainty in Artificial Intelligence (UAI)*, pages —. AUAI Press, 2017.
36. Joe Suzuki and Tianle Yang. Generalization of LiNGAM that allows confounding. In *2024 IEEE International Symposium on Information Theory (ISIT)*, pages 3540–3545. IEEE, 2024.
37. Thomas Verma and Judea Pearl. Causal networks: Semantics and expressiveness. In *Proceedings of the Fourth Annual Conference on Uncertainty in Artificial Intelligence (UAI-86)*, pages 69–76, Mountain View, CA, 1986. AUAI Press.

38. Y. Samuel Wang and Mathias Drton. Causal discovery with unobserved confounding and non-Gaussian data. *Journal of Machine Learning Research*, 24(79):1–62, 2023.
39. Kun Zhang, Jonas Peters, Dominik Janzing, and Bernhard Schölkopf. Kernel-based conditional independence test and application in causal discovery. In *Proceedings of the Twenty-Seventh Conference on Uncertainty in Artificial Intelligence (UAI 2011)*, pages 804–813. AUAI Press, 2011.

Index

J. Suzuki, *Graphical Models and Causal Discovery with Python*,
https://doi.org/10.1007/978-981-95-5308-2

MIX
Papier aus verantwortungsvollen Quellen
Paper from responsible sources
FSC® C105338

If you have any concerns about our products, you can contact us on
ProductSafety@springernature.com

In case Publisher is established outside the EU, the EU authorized representative is:
Springer Nature Customer Service Center GmbH
Europaplatz 3, 69115 Heidelberg, Germany

Printed by Libri Plureos GmbH
in Hamburg, Germany